THE EU AS INTERNATIONAL
ENVIRONMENTAL NEGOTIATOR

Global Environmental Governance

Series Editors: John J. Kirton, Munk Centre for Global Affairs, Trinity College, Canada and Miranda Schreurs, Freie Universität Berlin, Germany

Global Environmental Governance addresses the new generation of twenty-first century environmental problems and the challenges they pose for management and governance at the local, national, and global levels. Centred on the relationships among environmental change, economic forces, and political governance, the series explores the role of international institutions and instruments, national and sub-federal governments, private sector firms, scientists, and civil society, and provides a comprehensive body of progressive analyses on one of the world's most contentious international issues.

The EU as International Environmental Negotiator

TOM DELREUX
University of Louvain, Belgium

Routledge
Taylor & Francis Group

LONDON AND NEW YORK

First published 2011 by Ashgate Publishing

Published 2016 by Routledge
2 Park Square, Milton Park, Abingdon, Oxfordshire OX14 4RN
711 Third Avenue, New York, NY 10017, USA

First issued in paperback 2016

Routledge is an imprint of the Taylor & Francis Group, an informa business

British Library Cataloguing in Publication Data
Delreux, Tom.
   The EU as international environmental negotiator. – (Global environmental governance series)
   1. European Union – Foreign relations. 2. Environmental policy – European Union countries – Decision making.3. Environmental policy – International cooperation.
   I. Title II. Series
   363.7'0526'094–dc22

Library of Congress Cataloging-in-Publication Data
Delreux, Tom.
   The EU as international environmental negotiator / by Tom Delreux.
        p. cm.—(Global environmental governance)
   Includes index.
   ISBN 978-1-4094-1182-6 (hardback : alk. paper)—ISBN 978-1-4094-1183-3
1. Environmental policy—European Union countries. 2. European Union. 3. European Commission. 4. European Union countries—Foreign relations. 5. Negotiation—European Union countries. 6. Administrative procedure—European Union countries. I. Title.
   GE190.E85D45 2011
   363.7'0526—dc22

                                                            2010035119

ISBN 13: 978-1-138-27878-3 (pbk)
ISBN 13: 978-1-4094-1182-6 (hbk)

Contents

List of Figures

List of Tables

Acknowledgements

This book owes much to the support and input I got from a number of people and institutions. They all deserve my warmest thanks. First and foremost, I thank Bart Kerremans, with whom I worked together at the University of Leuven (K.U.Leuven). I learned a lot from his capacity to understand all kinds of political processes by exploring them from an analytical perspective. Our frequent talks about the study presented in this book, but also about Belgian, European and global politics in general, always gave me an intellectual boost.

This work has benefited greatly from the numerous suggestions, remarks and questions I received during discussions with other scholars. Various drafts of this work were carefully reviewed by Jan Beyers, Hans Bruyninckx, Peter Bursens, Chad Damro, Ana Maria Dobre, Edith Drieskens, Montserrat González Garibay, Stephan Keukeleire, Liselotte Libbrecht, Myriam Martins-Gistelinck, Christine Reh, Simon Schunz, Fried Swenden, Louise van Schaik, Peter Vermeersch, Gert Verschraegen and Joris Voets. The soundboard function they all fulfilled – each from his or her own point of view – was indispensable for arranging my ideas. Sara Dewachter, An Huybrechts and particularly Delphine Hesters helped me a lot with sharing their ideas about possible solutions for the methodological QCA obstacles I was facing. I also thank the anonymous reviewers of various papers and articles about the topic of this book for their helpful remarks.

The empirical fieldwork would not have been possible without the great contribution from various policy-makers who were closely involved in the negotiations of my research. All of the Commission and member state officials with whom I could freely talk about the European Union as a negotiator in international environmental negotiations largely contributed to my understanding of the political processes I was studying. I especially thank those people who not only generously answered my questions, but who also provided me further input and feedback and took the time to read through my case studies: Gerard Boere, Peter Boye, Didier Breyer, Dick de Bruijn, Jan De Mulder, Robert Donkers, Lieselotte Feldmann, Helmut Gaugitsch, Henri Jorritsma, Francesco La Camera, Jim Maund, Matthias Sauer, Helga Schrott, Gernot Schubert, Marc Smaers, Jan-Willem Sneep, Monica Törnlund, Willem Van Cotthem, Piet van der Meer, Mari Van Dreumel and Leena Ylä-Mononen. I am also very grateful to those officials who opened their personal archives to me and who offered me in this way a unique wealth of information.

My deep gratitude goes to Remy Merckx and Jan-Karel Kwisthout, who allowed me to become of member of various Belgian and Dutch delegations. They made it possible for me to experience in real life how decision-making processes at the European level and environmental negotiations at the international level work in practice. I am very thankful to all members of those delegations, who were constantly ready to guide me through these processes and to let me be fully submerged in the fascinating world of international environmental negotiations. This book would have looked fundamentally different if I had not been granted the opportunity to experience the subject of my research so intensely on the field.

I was able to conduct this study thanks to a one-year scholarship from the K.U.Leuven Research Council and a four-year fellowship from the Research Foundation Flanders ('FWO-Vlaanderen'). I am indebted to both institutions, which granted me the financial support to focus quasi-exclusively on this project. I also thank the 'Academische Stichting Leuven' for awarding me with a scholarship covering a part of the fieldwork.

At Ashgate, I thank Kirstin Howgate and Margaret Younger for their feedback and their continuous interest in the project, as well as the Global Environmental Governance series editors John Kirton and Miranda Schreurs.

My final thanks are for my non-academic environment. The ever lasting support and warmth from my family – both in Ieper and Koksijde – seemed self-evident, but was an indispensable certainty and backing when working on this study. Marjan deserves my thanks *hors catégorie*. Her love, support and understanding made writing this book so much easier.

Tom Delreux

List of Abbreviations

AETR	Accord Européen sur les Transports Routiers (European Road Transport Agreement)
AEWA	African-Eurasian Migratory Waterbirds Agreement
AIA	Advance Informed Agreement
AT	Austria
BE	Belgium
BSWG	Biosafety Working Group
CBD	Convention on Biological Diversity
CFSP	Common Foreign and Security Policy
CIS	Commonwealth of Independent States
CMS	Conservation of Migratory Species
COP	Conference of the Parties
Coreper	Comité des Représentents Permanents (Permanent Representatives Committee)
CSD	Commission on Sustainable Development
DE	Germany
DG	Directorate-General
DK	Denmark
DNA	Designated National Authority
EC	European Community
ECJ	European Court of Justice
EEC	European Economic Community
EIA	Environmental Impact Assessment
EL	Greece
EP	European Parliament
ES	Spain
EU	European Union
Euratom	European Atomic Energy Community
ExCOP	Extraordinary Meeting of the Conference of the Parties
FAO	Food and Agriculture Organization
FI	Finland
FR	France
FYROM	Former Yugoslav Republic of Macedonia
G77	Group of 77
GEF	Global Environment Facility
GMO	genetically modified organism
IE	Ireland
ILO	International Labour Organization

INC	Intergovernmental Negotiating Committee
INCD	Intergovernmental Negotiating Committee on Desertification
IT	Italy
Juscanz	Japan, United States, Canada, Australia, New Zealand
LMO	living modified organism
LRTAP	Long-Range Transboundary Air Pollution
LU	Luxembourg
MEA	multilateral environmental agreement
MEP	Member of the European Parliament
MOP	Meeting of the Parties
NGO	non-governmental organization
NL	The Netherlands
OECD	Organization for Economic Cooperation and Development
PIC	Prior Informed Consent
POPs	Persistent Organic Pollutants
PT	Portugal
QCA	Qualitative Comparative Analysis
QMV	qualified majority voting
RCI	rational choice institutionalism
REIO	Regional Economic Integration Organization
SB	Subsidiary Body
SEA (chap. 2)	Single European Act
SEA (chaps 4-6)	Strategic Environmental Assessment
SI	sociological institutionalism
SW	Sweden
TEC	Treaty establishing the European Community
TEU	Treaty on European Union
TFEU	Treaty on the Functioning of the European Union
UK	United Kingdom
UN	United Nations
UNCCD	United Nations Convention to Combat Desertification
UNCED	United Nations Conference on Environment and Development
UNECE	United Nations Economic Commission for Europe
UNEP	United Nations Environment Programme
UNFCCC	United Nations Framework Convention on Climate Change
US	United States
WEOG	Western European and Others Group
WG	Working Group
WPIEI	Working Party on International Environmental Issues
WTO	World Trade Organization

Chapter 1
Introduction

This book is about the European Union as a negotiator in international environmental negotiations. More in particular, it looks at the internal political processes in the EU when the EU negotiates multilateral environmental agreements (MEAs). In international environmental negotiations, the EU is mostly represented by a (set of) EU negotiator(s). The EU is often seen as a single actor, as a unitary negotiation partner. However, it is not. Within the EU, member states and the European Commission determine how the EU is functioning at the international negotiation level. These internal EU processes are analysed in this book. It focuses on the interplay between negotiations at the EU and at the international level. How does the EU function when it operates as global actor in the field of international environmental politics? How and to what extent do member states pool their voices, delegate negotiation authority and aim to speak with a single voice at the international level? In which way is a common EU position established? How does the institutional context of the EU transform the aggregated preferences of the member states into bargaining power that can be played in international negotiations? By answering such questions, this book analyses the decision-making process within the European Union in order to gain a better understanding of, on the one hand, the way the EU operates during international negotiations, and, on the other hand, the internal balance of power between the various EU actors.

The EU is usually seen as an important player in global environmental governance and international environmental negotiations. It has been labelled a 'frontrunner' (Damro, 2006), a 'crucial actor' (Damro, Luaces, 2001), an 'influential global actor' (Rhinard, Kaeding, 2006), or a 'powerful global actor' (Jokela, 2002). Some scholars even portray the EU as an 'environmental leader' (Zito, 2005; Kelemen, 2010).[1] Because the EU is an important player in the

---

1   The claim that the EU shows leadership in international environmental negotiations is particularly made with regard to biotechnology (Tiberghien, Starrs, 2004; Falkner, 2007) and climate change (Yamin, 1998; Chagas, 2003; Cameron, 2004; Costa, 2006; Schreurs, Tiberghien, 2007; Oberthür, Roche Kelly, 2008; Schmidt, 2008). However, other scholars question the leadership role of the EU, stating that this claim is only made by the EU itself (Jokela, 2002; Bretherton, Vogler, 2003; Dessai, Schipper, 2003; Vogler, 2005), that the EU has not yet fulfilled its leadership potential (Liberatore, 1997; Ott, 2001), that the EU is not perceived as a leader by the negotiation partners (Elgström, 2006; Elgström, 2007), or even that the EU is characterized by a 'leadership deficiency' (Sjöstedt, 1998). Although scholars do not agree on the leadership character of the EU, there is a broad agreement in the literature that the EU is a key player in international environmental negotiations.

negotiations, the negotiation role, positions, strategy and bargaining power of the EU have to be taken into account in order to understand the process and the outcome of international negotiations leading to an MEA. To comprehend the political dynamics and events at the international level, understanding the processes at the level of the key players is important. Therefore, the political processes inside the EU need to be unravelled. In the current international relations literature, it is generally assumed that explaining the behaviour of a political actor negotiating at the international level requires taking into account the politics at its internal level (Moravcsik, 1993; Milner, 1997; Jupille, 1999). Such actors are not black boxes and their internal political dynamics matter. This claim is certainly true when analysing the EU as an international actor. The EU is not a state with a single domestic level, but a complex multi-level polity with its own institutional context in which the preferences of the future parties to the MEA are aggregated, and possibly formed and transformed.

In conceptual terms, two kinds of actors take a central place in EU decision-making processes with regard to international negotiations: member states and EU negotiators. The member states are often represented by one or more actors, negotiating on their behalf. I label these actors the 'EU negotiator(s)'. In practice, the EU negotiator is usually the Commission, the Presidency, one or more member state(s), or any combination of these. The EU decision-making processes studied in this book are thus characterized by the relation between actors who are represented (the member states) on the one hand, and actors who represent them (the EU negotiators) on the other hand.

The central concept, around which the analysis in this book is built, is 'discretion'. Discretion should be understood as the autonomy, the range of potential independent action or the degree of freedom enjoyed by the EU negotiator vis-à-vis the member states. The discretion enjoyed by the EU negotiator vis-à-vis the member states is crucial to understand the dynamics and the outcomes of the international negotiations. It determines the EU's flexibility with regard to reacting on proposals by the negotiation partners and it determines how far the EU negotiator can go along with the direction in which the international negotiation process and its outcome are moving. In other words, the EU negotiator's discretion defines whether the EU negotiator constantly needs to consult the member states or whether the EU negotiator can react with a certain degree of autonomy to the dynamics of the international negotiations. The extent to which the EU negotiator's hands are tied is thus decisive for his negotiation behaviour at the international level (Moravcsik, 1993; Meunier, 2005; Meunier, Nicolaïdis, 2006).

1. Aim of the book

1.1. Research question

The aim of this book is to answer the following research question: *which conditions determine the EU negotiator's discretion vis-à-vis the member states during international negotiations leading to a multilateral environmental agreement?* In order to identify the conditions under which the EU negotiator enjoys a particular degree of discretion, this book goes beyond the single case study method, as it has been used in studies on the EU decision-making process with regard to other external policies of the EU (mostly in the trade policy domain). It analyses and compares the EU negotiator's discretion in eight international negotiations leading to an MEA: the UN Desertification Convention, the African-Eurasian Waterbirds Agreement, the Kyoto Protocol, the Aarhus Convention, the Rotterdam PIC Convention, the Cartagena Protocol, the Stockholm POPs Convention and the SEA Protocol.

By studying the EU negotiator's discretion, I focus on the core element of the EU decision-making process and of the relation between the representing and the represented actors: the balance of power in the EU. Do the member states still dominate the process or is the degree of discretion enjoyed by the EU negotiator so high that the latter rules the process? Which actors or which institutions are pulling the strings in the context of the EU as an actor in international environmental politics: those who delegate or those who represent, or – in the theoretical terms used in this study – the principals or the agent? In their overview and assessment of research dealing with similar questions in the area of the EU's external trade policy, Dür and Zimmermann mention that various authors come to different conclusions on the question whether the EU negotiator enjoys a high degree of discretion or not (Dür, Zimmermann, 2007).[2] This may indicate that the EU negotiator's discretion is not constant in every decision-making process and that the research focus should be shifted from explaining the EU negotiator's discretion in a single decision-making process towards identifying the scope conditions and causal mechanisms of discretion. This book takes the observation by Dür and Zimmermann as a starting point and aims to identify the conditions under which the EU negotiator enjoys a particular degree of discretion.

1.2. Puzzle

Studying EU decision-making processes with regard to negotiations leading to an MEA not only contributes to a better understanding of the EU's role in

---

2 As examples of studies demonstrating that the Commission dominates the process, Dür and Zimmermann mention Johnson, 1998; Coleman, Tangermann, 1999; and Woll, 2006. By contrast, examples of studies emphasizing member state dominance are Aggarwal, Fogarty, 2004; De Bièvre, Dür, 2005; Meunier, 2005; and van den Hoven, 2004.

international environmental politics. It is also particularly relevant and interesting from the perspective of EU politics because of a twofold reason. First, it teaches us something about the way the EU deals with so-called 'mixed agreements', which are an increasingly important component of the EU's external relations. Second, the study of these decision-making processes fits in the framework of the general debate about the balance of power in the EU. Both the member states and the EU negotiators are faced with dilemmas. Their respective trade-offs do not only determine the way the EU negotiates at the international level, but also the internal balance of power within the EU. Let us have a deeper look at these two points.

1.2.1. Mixed agreements as an increasingly important component of the EU's external relations

Most research on the EU as an international actor deals with either the Common Foreign and Security Policy (CFSP) or the EU's external trade policy. These policy areas constitute the ends of the intergovernmental-supranational continuum of the EU's external relations. CFSP is characterized by an intergovernmental dynamic, nationalcompetencesandmemberstatesmaintainingtheirsovereigntyasinternational actor, although possibly represented by a common representative, such as the High Representative or previously the rotating Presidency. External trade policy, by contrast, has a more supranational character, based on EU competences, which mostly results in the Commission negotiating at the international level on behalf of the member states. While the CFSP is mainly member state driven, the Commission plays a key role in trade policy-making. The EU's external environmental policy is situated somewhat in-between both ends of the continuum.

So far, less scholarly attention has been paid to what happens in the middle of this continuum. This might surprise as many issues in international politics are covered by the competences of both the EU and the member states. These are 'shared competences' and the international treaties touching upon this kind of competences are 'mixed agreements'. Both the EU and the member states are a party to these agreements. Mixed agreements do not only occur in the environmental policy domain. Other well-known examples are the EU-US Open Skies Agreement, various Stabilization and Association Agreements with Balkan countries, or trade agreements including the so-called 'new trade issues'[3] (Meunier, Nicolaïdis, 2000; Leal-Arcas, 2003; De Bièvre, Dür, 2005; Billiet, 2006; Dür, Zimmermann, 2007). Nowadays, mixed agreements are thus part and

---

3   In the introduction to the special issue of Journal of Common Market Studies on the EU in international trade negotiations, Dür and Zimmermann mention that 'the constant expansion of the scope of international trade created substantial tensions between the Council of Ministers and the Commission over how to negotiate and how to ratify international agreements on "behind the border" issues' (Dür, Zimmermann, 2007, pp. 773-774). This puzzle, emerging the last decade in the international trade area, has been present in the EU decision-making processes with regard to international environmental negotiations since a long time.

parcel of the day-to-day practice in the external relations of the EU. Therefore, the limited attention in the political science literature for the way the European Union copes with (negotiations on) mixed agreements is striking. By contributing to the understanding of the functioning of the EU in this kind of international politics, this book aims to fill a significant gap in the existing literature.

As both the EU and the member states will become a party to an MEA, they are both present around the international negotiation table. The issues covered by EU competences should be negotiated by the European Commission, while the member states can formally negotiate the issues on which they still have competences. However, such a clear division between EU and national competences undermines the EU's ability to speak with a single voice, which is assumed to decrease its bargaining power (Frieden, 2004). Practice teaches us that the member states usually opt to pool their voices and to delegate negotiation authority to one or more EU negotiator(s). Question is then how the EU deals with the tension between the striving to be perceived as a strong negotiation bloc on the one hand and the legal situation, which suggest a division in the negotiation arrangement, on the other hand (Delreux, 2006). The EU decision-making process with regard to negotiations leading to a (mixed) MEA is a clear example of the hybrid nature of the EU: neither a state nor an international organization, neither completely supranational nor fully intergovernmental. It is intriguing to examine how the EU deals with this tension between supranationalism and intergovernmentalism, given the particular political international context of international environmental politics in which the EU strives to play an important role. In this respect, studying these processes goes to the heart of the EU's nature.

1.2.2. Balance of power in the EU: a dilemma for the member states and the EU negotiator

When MEAs are negotiated, both the member states and the EU negotiator are confronted with a political dilemma. This means that they have to make choices and that these choices determine their political behaviour in the EU decision-making process. One of the driving questions behind the study of the EU negotiating MEAs is how the member states and the EU negotiator cope with their dilemmas.

The dilemma for the member states departs from the assumption that member states want to influence MEAs. This is a plausible assumption because these international treaties affect to a large extent the costs and benefits of the domestic policies of their parties, including the EU member states. They need to be implemented, which influences policies in various domains, including agricultural, industrial, transport, external trade or food safety policy. Because of the deep and widespread effect on their domestic policies and political choices, national governments seek to influence the outcome of the international negotiations.

Aiming to influence the substance of an MEA, member states are confronted with a dilemma. On the one hand, a member state has to make sure that its preferences about the provisions of the MEA are presented at the international level, thereby aiming that the outcome of the negotiations moves in the direction of that preference. This strategy requires that the preference of a member state is purely and clearly presented during the negotiation process. On the other hand, a member state has to make sure that its preference is presented as strongly as possible at the international level and that the negotiation partners take it into account. This strategy requires that the preference is backed by a large degree of bargaining power. In the context of the EU, this can be achieved by working via an EU negotiator, representing the member states jointly.

However, these two strategies are not fully compatible, which generates a dilemma for the member states. The first strategy is most adequately realized by negotiating on its own, without being hindered by the straitjacket of the EU and the common position resulting from the aggregation of the various member state preferences. Indeed, the aggregation of preferences with the other member states by definition entails compromising on one's own preference. To realize the second strategy, however, it is instrumental to develop a common EU position to be presented at the international level on behalf of the EU as a whole. It is clear that the cost of the one strategy is the benefit of the other, and vice-versa. The benefit of negotiating directly at the international level is that a member state is sure that its own preference is presented during the international negotiations, while its cost is that the relative bargaining power behind that preference may be very limited. This can be undone by negotiating via the EU and by delegating negotiation authority to an EU negotiator (benefit of negotiating via the EU level), although this implies that its preference can be distorted with these of the other EU actors (cost of negotiating via the EU level).

This dilemma and the constant trade-off that member states have to make is the key puzzle when looking at the EU decision-making process from the member states' point of view. Do the member states grant a large degree of autonomy to an actor who speaks on their behalf, do they limit his discretion or do they even preserve this power for themselves, given the costs and benefits of delegation to an EU negotiator? To what extent are national governments prepared to yield authority and power for negotiations that will affect the national policies they will have to execute and to sell to their constituents afterwards?

Not only the member states are confronted with a dilemma about the degree of negotiation autonomy they should grant to the EU negotiator. Also the EU negotiator faces a dilemma about the discretion he enjoys. Since the EU negotiator operates simultaneously at the international and the EU level, he is engaged in two negotiation contexts, generating different – and often contradictory – expectations about his behaviour. On the one hand, the EU negotiator has to represent the member states and their common position, which results from an aggregation of the various national preferences. In that sense, the EU negotiator's hands are tied by the member states (Meunier, 2005; Meunier, Nicolaïdis, 2006), although the degree of being bound can considerably vary. What matters is indeed that the EU negotiator ultimately has to get the approval of the member states on the internationally reached agreement. On the other hand, the EU negotiator is involved in the negotiation dynamics at the international level. At this level, the

EU negotiator experiences the pressure coming from negotiation partners to agree – and thus to compromise – on certain provisions of the MEA. Consequently, the EU negotiator faces a Janus-like role, as he is the actor connecting two political processes at different levels, which generate different expectations from him (Putnam, 1988; Pattersson, 1997; Damro, Hardie, MacKenzie, 2008).

The key question is how far the EU negotiator is able to engage in the international negotiation process, while still reasonably expecting that the member states will not reject his international commitments afterwards. The delicate trade-off that he constantly has to make is to satisfy the international negotiation partners by making commitments in order to reach an international agreement, while assuring that the member states will not blow the whistle on him afterwards. How much additional autonomy can the EU negotiator conquer vis-à-vis the member states to negotiate at the international level until he will face a rejection at the EU level, causing a loss of face vis-à-vis the international negotiation partners?

1.3. Argument and data

This book sheds light on the EU decision-making process with regard to international environmental negotiations and aims to explain the discretion enjoyed by the EU negotiator in such processes. On the basis of comprehensive case studies and a formalized cross-case analysis, the following claims are made:

- the EU decision-making process can be modelled as a principal–agent relation that is extended in a threefold way: taking into account the international negotiation context, considering the mixed nature of MEAs, and incorporating dynamics that emphasize the possible cooperative nature of EU decision-making;
- the compellingness of the international negotiations, i.e. the pressure coming from the external negotiation partners to reach agreement, is a necessary condition for an EU negotiator enjoying discretion vis-à-vis the member states. However, more qualified explanations, stressing the role of institutional density, the level of politicization and the distribution of preferences and information, are needed to understand why the EU negotiator enjoys a particular degree of discretion;
- the relation between member states and their negotiator is characterized by an interplay of bidirectional cooperation and control;
- in practice, the EU decision-making process does not completely follow the path that can be expected from a legal point of view. Pragmatism is often more decisive than formal rules and the division of competences, both in determining how the EU negotiation arrangement is organized and how the decision-making process develops; and
- both the member states and the EU negotiator are able to find a balance in their respective dilemmas as sketched above, if they invest sufficient and appropriate political capabilities in the process.

The empirical data for this study is based on different and multiple sources. This triangulation allows me to strive for confirmation and completeness (Arksey, Knight, 1999): the different sources are used to check each other's reliability (confirmation) and to solve each other's shortcomings (completeness). Moreover, it allows me to apply the data collection strategy of 'process-tracing': I reconstruct the (decision-making) processes on the basis of various sources (Checkel, 2001; Pollack, 2002; Checkel, 2004; Niemann, 2004; Checkel, 2005a; Checkel, 2006). Five different sources of information are used.

First, the main source of information was *semi-structured interviews* that were conducted with officials from the member states and the Commission who were involved in the decision-making processes.[4] My questions probed the respondents' evaluation of a certain decision-making situation. The questions did not deal with the interviewees' own positions or decision-making behaviour: they rather measured their assessment of the decision-making process as a whole.[5]

Second, *secondary literature* on the studied cases was consulted. Although there is a high degree of variation in the number of publications about the various EU decision-making processes and international environmental negotiations, a couple of articles were available on most of the cases.[6]

Third, *media and press reports*, covering the international negotiations, were checked. For information on the negotiations at the international level, *Earth Negotiation Bulletin* is an indispensable tool, since this online publication covers, on a daily basis, most global environmental negotiations.[7] For following the decision-making process at the EU level, I used *Agence Europe* and *European Voice* as media sources.[8]

---

4   All interviews were conducted between June 2006 and January 2007. The affiliation of the 61 interviewees, their role in the decision-making process, and the details of the interview are included in appendix A.

5   In order to check whether my interpretation of the data corresponded to the reality of these decision-making processes, each of the case studies presented in Chapter 4 was read and reviewed by a couple of interviewees, who proposed corrections and supplementations if this was needed. This check allows me to claim a high degree of validity for my empirical data.

6   However, as the majority of these articles covered the substance, consequences or an evaluation of the MEAs, very few of them dealt with the negotiation process as such or with the role of the EU during these negotiations.

7   *Earth Negotiation Bulletin* is published by the International Institute for Sustainable Development. Its archive is available on http://www.iisd.ca/voltoc.html. From the negotiations that were studied in this research, the negotiations on UNCCD, the Kyoto Protocol, the Rotterdam Convention, the Cartagena Protocol and the Stockholm Convention were covered by *Earth Negotiation Bulletin*. The three 'less multilateral' negotiations (on AEWA, on the Aarhus Convention and on the SEA Protocol) were not covered.

8   However, the level of attention paid to the selected EU decision-making process was rather limited, even in these publications.

Fourth, I had the opportunity to investigate *informal and (semi-)confidential documents* with regard to the EU decision-making processes. Official EU documents on the studied decision-making processes only contain the official decisions to sign or to ratify the MEA. Neither the official reports of the Council meetings nor the official press releases provide detailed information on the variables used in this study. The real interesting information is often preserved in decision-makers' personal archives and informal, (semi-)confidential documents. For five cases, I could examine the personal archives of some of my interviewees. In this way, I collected personal notes and reports on the various negotiation sessions and the EU decision-making processes on the one hand, and confidential EU documents and notes on the other hand.

Fifth, *participatory observation* in both international environmental negotiations and EU decision-making processes with regard to these negotiations made my views and ideas about the EU as international environmental negotiator more concrete and it helped to contextualize my findings.[9]

2. Structure of the book

In order to answer the research question, a number of successive steps are taken. I begin with a legal perspective on the EU decision-making processes with regard to international environmental negotiations (Chapter 2). Starting from the research subject and research question, I then elaborate a theoretical model, on the basis of which the research design is developed (Chapter 3). The empirical part of this book consists of one descriptive and two analytical chapters (Chapters 4-6). Finally, the conclusion answers the research question and refers back to the theoretical insights (Chapter 7).[10]

---

9   As the selected cases all took place before my research started, I was not able to attend those EU decision-making processes. However, to get an insight in the structure and the dynamics of international environmental negotiations and EU decision-making processes with regard to them, I attended similar meetings, both at the EU and the international level.

10   Since some chapters draw on ideas that were previously published, I thank the respective publishers for allowing me to use these materials. Chapter 2 is based on my article 'The European Union in international environmental negotiations: a legal perspective on the internal decision-making process', *International Environmental Agreements*, 6(3), 2006, 231-248. The first part of Chapter 3 makes use of material first published in 'The EU as a negotiator in multilateral chemicals negotiations: multiple principals, different agents', *Journal of European Public Policy*, 15(7), 2008, 1069-1086. Sections of Chapter 4 have appeared in 'The European Union in International Environmental Negotiations: an Analysis of the Stockholm Convention Negotiations' in *Environmental Policy and Governance*, 19(1), 2009, 21-31 and in 'The EU in Environmental Negotiations in UNECE: An Analysis of its Role in the Aarhus Convention and the SEA Protocol Negotiations' in *Review of European Community and International Environmental Law*, 18(3), 2009, 328-337.

The current Chapter 1 revealed why the EU decision-making process with regard to international negotiations leading to an MEA is a relevant and challenging research topic. Studying this topic promises to provide interesting insights about both the processes in international environmental politics and about the political dynamics at the EU level. On the basis of these puzzles and the gaps in the existing literature, the research question was defined, emphasizing the quest for the conditions determining discretion enjoyed by the EU negotiator.

Chapter 2 describes the legal framework of the EU decision-making process with regard to mixed agreements. As a legal framework determines the basic rules of the game on the basis of which political actors engage in political processes, I start the examination of the EU negotiator's discretion by looking at what the Treaties and the rulings by the European Court of Justice (ECJ) say about the EU conducting international negotiations resulting in mixed agreements.

Chapter 3 translates the elements from the legal decision-making procedure into a theoretical framework. As the decision-making process is essentially about the delegation of negotiation authority from one set of political actors to another, I use the principal–agent model to theoretically frame the decision-making process. The member states are conceptualized as 'principals', the EU negotiator as 'agent'. After building the principal–agent model, I extend it in a threefold manner, allowing me to introduce the particular reality of international environmental negotiations into the model. First, the concepts 'private information for the principals', 'political cost of no agreement' and 'compellingness of the external environment' are introduced, and their expected effects on the agent's discretion are discussed. Second, I extend the model to grasp the mixed nature of MEAs. Third, I enrich the model with insights from sociological institutionalism. At the end of Chapter 3, I present the research design. On the one hand, the variables are selected in a theory-driven way. On the other hand, I argue why I analyse the eight selected cases: the EU decision-making processes with regard to the international negotiations leading to the UN Desertification Convention, the African-Eurasian Waterbirds Agreement, the Kyoto Protocol, the Aarhus Convention, the Rotterdam PIC Convention, the Cartagena Protocol, the Stockholm POPs Convention and the SEA Protocol.

Chapter 4 is the descriptive part of this book. The different aspects of the eight selected EU decision-making processes are presented. This chapter offers a descriptive and systematically structured overview of the cases. It provides the reader with eight stories of how different EU decision-making processes with regard to negotiations leading to MEAs worked in practice. Moreover, this chapter functions as a point of reference to consult the empirical data when the next analytical chapters are read.

Chapter 5 expands the findings published in 'The EU negotiates multilateral environmental agreements: explaining the agent's discretion' in *Journal of European Public Policy*, 16(5), 2009, 719-737, and some sections in Chapter 6 are based on 'Cooperation and Control in the European Union. The Case of the European Union as International Environmental Negotiator' in *Cooperation and Conflict*, 44(2), 2009, 189-208.

In Chapter 5, the data is analysed by means of Qualitative Comparative Analysis (QCA). By guiding the reader through the different steps of the empirical analysis, I show which combinations of conditions explain a particular degree of discretion enjoyed by an EU negotiator. Moreover, the QCA results are interpreted in the light of the empirical data, presented in the previous chapter, and the causal mechanisms behind the combinations of conditions and the outcome are identified. This allows me to combine the explanatory power of QCA with the empirical richness of my data.

Chapter 6 analyses the EU decision-making process with regard to international negotiations leading to an MEA from a more general point of view. Whereas the previous chapter took the case-specific characteristics of the decision-making processes into account, this chapter aims to make more generalizable claims about the way such EU decision-making processes work in practice. Therefore, this chapter returns to the variables. Going back and forth between theory and empirics, this chapter discusses the general organization and determinants of the EU negotiation arrangement and the various aspects in the relations among the actors involved in the decision-making process. I also evaluate the different steps and the outcomes of the decision-making process and I address the extent to which sociological institutionalist elements have to be taken into account to supplement the traditional rational choice inspired principal–agent perspective on delegation from the member states to an EU negotiator.

In Chapter 7, the conclusions of this book are presented. Taking into account the results of the empirical analysis, I return to each of the previous chapters and I formulate answers to the questions developed in these chapters. More in particular, I answer the research question, as presented in the current chapter, and I elaborate how the member states and the EU negotiator cope with the dilemmas identified above. Furthermore, the legal framework of the decision-making process is compared to the way the process is conducted in practice. The theoretical framework – both the basic principal–agent model and the three extensions – is evaluated and the assumptions and propositions are assessed. I also evaluate the QCA research approach.

Chapter 2
The EU and Mixed Agreements

1. Introduction: the legal framework as a starting point

This chapter describes the legal framework within which the EU decision-making process with regard to international negotiations leading to an MEA takes place. Decision-making processes in the EU do not occur in a legal vacuum, as primary law in the Treaties and various rulings by the ECJ constitute the rules of the game, which determine the borders of the political decision-making process. By taking the legal framework as a starting point for the analysis, this chapter meets the critique that 'the legal dimension of principal–agent relations is often under-articulated in studies of EU policy-making' (Maher, Billiet, Hodson, 2009, p. 412).

Depicting the legal framework within which the EU negotiates MEAs confronts us with an additional difficulty: the current legal framework (under the Treaty of Lisbon) is not the same as the legal framework determining the EU decision-making processes studied in this book, which date from before the Lisbon Treaty entered into force, and which were thus conducted under the Treaties of Nice, Amsterdam or Maastricht. Since the empirical cases presented in this book all took place in the period 1993-2004, the legal provisions that are sketched in this chapter are the ones that were in force in the pre-Lisbon era. These pre-Lisbon provisions are appropriate to understand the empirical cases that are analysed in the next chapters. However, although the current chapter primarily discusses the pre-Lisbon framework, it also mentions the new elements and dynamics on the EU's external environmental policy that are introduced in the Lisbon Treaty.

As will become clear from this chapter, the legal boundaries of the EU decision-making processes with regard to the negotiation of mixed agreements leave room for political interpretation. The procedure for negotiating MEAs is only partially regulated in the EU and there is no all-embracing legal framework. The reason is that the issues of an MEA are covered by both EC competences[1] and national competences. Consequently, neither European legal provisions nor national legal provisions can regulate these processes completely. This creates room for political flexibility, which largely determines the decision-making process and the relation between the various actors engaging in this process. While the next chapters

---

1   This is already a first element that has been modified by the Lisbon Treaty: before the entry into force of the Lisbon Treaty, it was the European *Community* that possessed competences at the European level, whilst since the Lisbon Treaty it is the European *Union* having these competences. The reason is that the Community is no longer a legal entity under the Lisbon Treaty.

focus on the way in which political actors fill in this political room, this chapter elaborates on its legal boundaries.

Section 2 lists the conditions the EU needs to fulfil for participating in international environmental negotiations and for concluding an MEA. It particularly focuses on the division of competences on environmental issues, because this is the main determinant of the legal framework. Since environmental competences are shared between the EC and the member states,[2] I distinguish between a discussion of the EC part of the process (Section 3), and a discussion of the member states part (Section 4). Finally, Section 5 brings both parts together and it compares the legal provisions of the decision-making process dealing with shared competences with those regulating the decision-making process touching upon exclusive EC competences. This comparison allows me to identify the unique points of mixed agreements, being the result of shared competences, and their consequences for the EU decision-making process.

2. Conditions for the European Union to act in international environmental negotiations

Like other international entities, the EU has to fulfil three necessary conditions to be a subject of international law and to participate in international environmental negotiations as a future party to the MEA: possessing legal personality, being recognized by its negotiation partners, and having the necessary competences to make binding commitments.[3] In this section, these three conditions are checked for the EU as a negotiator in multilateral environmental negotiations.

2.1. International legal personality

A first condition to conclude international agreements and to become a party to these is possessing international legal personality. This refers to an actor's capacity to act in the international system with legal effects. The answer to the question whether the EU possesses the necessary international legal personality to act internationally in the field of environmental policies has changed since the coming into force of the Lisbon Treaty. Indeed, before December 2009, the European Community had international legal personality, whereas nowadays the

---

2   To correspond with post-Lisbon provisions, also in this paragraph, 'EU' should be used instead of 'EC'.

3   These three conditions also relate to the EU's actorness, i.e. the extent to which the EU is an actor in international politics or negotiations, since they are three of the four criteria (or dimensions) of actorness (Jupille, Caporaso, 1998; Vogler, 1999; Groenleer, van Schaik, 2007). The fourth actorness criterion is the EU's autonomy vis-à-vis the member states, which is the central concept of this book.

European Union has gained this status. The replacement of the EC by the EU was indeed one of the innovations introduced by the Treaty of Lisbon.

During the period studied in this book, and from the perspective of international law, the European Community met the criteria for international legal personality, as specified in the 1949 International Court of Justice case Reparation for Injuries. The legal personality of the EC was also confirmed by article 281 TEC, stating that '[t]he Community shall have legal personality', as well as by the ECJ Costa/ENEL case (Verwey, 2004). Consequently, before the entry into force of the Lisbon Treaty, the EC was able to conclude international (environmental) agreements (Emiliou, 1996).

2.2. External recognition

Possessing international legal personality is not sufficient for the EC/EU to fully participate in international negotiations and to become a party to international agreements. A second condition is to be recognized by the other parties as a negotiation partner and as a future party to the agreement (Macrory, Hession, 1996; Jupille, Caporaso, 1998; Vogler, 1999). In the 1970s and 80s, the path towards external recognition in international environmental politics has been a struggle for the EC. The complicated and evolving division of competences, the lack of precedents, and the uncertainty this generated among the negotiation partners are considered as the main causes for this struggle (Sbragia, 1998). However, as the recognition question is not really an issue anymore in today's international negotiations, it seems that this struggle has been won by the EC/EU (Vogler, 1999). Nowadays, the EC/EU is widely recognized as a possible party to MEAs. It is usually recognized as a REIO, which stands for 'Regional Economic Integration Organization'. Although this concept can formally be used to recognize a variety of such organizations, it has only been applied to the EC/EU until now.

2.3. Competences

Finally, besides international legal personality and a recognized status, the EC/EU also has to hold competences in the field of the negotiated issues. Since competences were attributed to the EC in the period covered in this book, I use the term 'EC competences' rather than 'EU competences', as the latter should only be used when analysing questions on competences after the entry into force of the Lisbon Treaty. The competences condition is the most complicated one and needs a more fine-tuned and detailed elaboration. In what follows, I divide the competence condition in two subquestions. First, did the EC have external competences to become a party to MEAs, and, if so, where did these competences originate from (Section 2.3.1)? Second, if the first question reveals that the EC had competences to act in international environmental negotiations, were these competences then exclusively attributed to the EC (Section 2.3.2.)?

2.3.1. External competences?

The attribution of external competences to the EC has been realized in two ways: by express powers and by implied powers.

Express powers are competences explicitly mentioned in the TEC, or nowadays in the TFEU. The Treaty of Rome (1957) only contained two policy fields with express external competences for the European Economic Community (EEC): trade policy and the conclusion of association agreements with third states. The Single European Act (SEA, 1986-1987) added express powers in the field of external environmental policy (Jupille, Caporaso, 1998; Peterson, Bomberg, 1999). Under the Treaty of Nice, the range of policy areas with express external EC competences increased, as it also included monetary, research and development policy.

At the same time, external competences of the EC broadened in a more indirect way, namely by implied powers. Implied powers are the actualization of Treaty provisions, established by a legal doctrine by the ECJ. Indeed, the ECJ developed the principle that EC competences could also be derived from the interpretation of the Treaty (Macleod, Hendry, Hyett, 1998; Leal-Arcas, 2001). The rationale behind the establishment of implied powers is the doctrine of parallelism between internal and external competences. Broadly speaking, this doctrine stipulates that internal and external competences need to be closely connected. During the 70s, four main cases of the ECJ broadened the scope of the EC's external competences by developing implied powers. First – and most far-reaching –, the Case 22/70 (AETR) can be considered the starting point for the implied powers. In the AETR case, the ECJ stated that when the EC had taken measures to realize a common policy, the member states were no longer allowed to conclude international agreements that (could) undermine these internal EC measures (Dashwood, Heliskoski, 2000; Pocar, 2002; Verwey, 2004). AETR thus created the doctrine of parallelism between internal and external competences: if the EC had elaborated measures in a particular policy area, it could also conduct external relations in that domain and it could become a party to international agreements covering that domain (Sbragia, 1998). Following this 'in foro interno, in foro externo' principle, conducting external policy does not have to be based on explicit provisions in the Treaty: it can also arise from secondary legislation (Leefmans, 1998).

Second, the Kramer Case confirmed the AETR argument that external powers can also be adopted via internal legislation (Leefmans, 1998; Dutzler, 2002; Verwey, 2004). Both AETR and Kramer entail that the internal powers, on which the external competences can be based, have to be exercised before the parallelism doctrine can be applied.

Third, the parallelism principle was expanded by Opinion 1/76 (Rhine Navigation). The ECJ stated that an external competence can be exercised on the basis of internal competences, even if the internal competences have not been exercised before. However, this is only possible under the condition that the participation of the EC in an international agreement is necessary to realize the internal objectives mentioned in the Treaty (Leefmans, 1998; Macleod, Hendry, Hyett, 1998; Dashwood, Heliskoski, 2000). This means that implied external competences can be created even without the existence of secondary EC legislation, as far as it is necessary to realize a Treaty objective.

Fourth, in Opinion 1/78 (Natural Rubber), the ECJ created external competences that are not derived from internal competences. It reasoned that, when the main issues of an international agreement fell under the competences of the EC and the side-issues did not, the EC could act in the area of the side-issues if this occurred in close connection with the main issues (Leefmans, 1998).

Subsequently, the principle of implied powers and the doctrine of parallelism has been repeated and emphasized by the ECJ, for instance in Opinion 2/91 and in Opinion 1/94.

2.3.2. Exclusive external competences?

Both internal and external competences can be exclusively or non-exclusively attributed to the EC. In policy areas with exclusive competences, member states are no longer allowed to act on their own (Heliskoski, 2000; Leal-Arcas, 2001; Eeckhout, 2004). Non-exclusive competences, by contrast, are shared competences, on which both the EC and the member states are still allowed to act.

In the field of external environmental policy, EC competences are most of the time not exclusive. Environmental competences are thus shared competences. Two TEC articles guaranteed that the member states preserve the necessary competences to act internationally on environmental issues and that competences were non-exclusively attributed to the EC. One article mentioned this explicitly, another implicitly. First, the non-exclusivity of EC external environmental competences was pointed out in article 174§4 TEC (nowadays article 191§4 TFEU),[4] which stipulated that the EC and the member states can internationally act in their respective spheres of competences, as this occurs 'without prejudice to Member States' competence to negotiate in international bodies and to conclude international agreements'. Second, article 176 TEC (currently article 193 TFEU) prescribed that policy measures adopted by the EC 'will not prevent any Member State from maintaining or introducing more stringent measures'. In other words, even if environmental standards are harmonized within the EU, member states are allowed – also at the international level – to take measures that are even more far-reaching.

The division of competences between the EC and the member states is the main determining factor for the nature of international agreements from a European perspective. While agreements touching upon issues on which the EC has exclusive competences are only negotiated and concluded by the EC, agreements on issues on which both the EC and the member states have competences are negotiated and concluded by the EC and the member states. These are mixed agreements. Consequently, two kinds of international agreements are possible: exclusive EC competences lead to an exclusive EC agreement and shared competences lead

---

4   The Lisbon Treaty does not introduce major changes to the environment articles in the TEC. Hence, articles 191-193 TFEU are taken up from articles 174-176 TEC.

to a mixed agreement. Although shared competences are nowadays rather the rule than the exception in the EU's external relations – particularly in the field of environmental policy –, the negotiation of mixed agreements *in se* is not legally regulated in the EU.

Mixed agreements thus cover both EC and national competences. Clarifying the division of competences within the EU is, however, a difficult task (Heliskoski, 2001; Eeckhout, 2004). In international environmental negotiations, it is often difficult to draw the line between issues falling under EC competence and issues covered by the competences of the member states (Emiliou, 1996; Vogler, 1999; Eeckhout, 2004). Each time international environmental negotiations are organized, the EU is confronted with new questions on the division of competences because of the evolution in the internal legislation on environmental matters.

A mixed agreement can be defined as an international agreement to which both an international organization (*in casu* the EC/EU) and one or more of its member states are parties (Leefmans, 1998; Macleod, Hendry, Hyett, 1998; Leal-Arcas, 2001; Thieme, 2001; Betz, 2008). Shared competences are the main reason for an international agreement to be mixed (Koutrakos, 2006). Indeed, when both the EC and the member state have competences on the issues regulated in the agreement and when they agree on the text, they both have to become a party to it.

Since MEAs include an EC part and a member states part, one could – at least theoretically – distinguish between two parallel decision-making processes (McGoldrick, 1997; Heliskoski, 2001). On the one hand, the framework regulating the relations between the EU actors for issues falling under EC competences were stipulated in article 300 TEC (replaced by article 218 TFEU since December 2009), which I discuss in the next section. On the other hand, member states may opt to pool their preferences and voices. If they do so, they set up an *ad hoc* negotiating mechanism. The legal framework of this member state part, which is not completely regulated in any European Treaty, is discussed in Section 4.

3. The EC as external environmental negotiator

To conclude MEAs, the EC needs a twofold legal basis. On the one hand, article 300 TEC – currently article 218 TFEU – functions as the general article regulating the *procedural* aspects for EC decision-making processes with regard to issues on which the EC has competences. On the other hand, a reference to the *substantive* competence is needed, since article 300 TEC/218 TFEU only regulates the procedure and since it does not give the permission to act externally (Macleod, Hendry, Hyett, 1998; Lenaerts, Van Nuffel, Bray, 2005). For international environmental matters, this substantive legal basis could be found in articles 174 and 175 TEC, which are currently articles 191 and 192 TFEU. For matters of consistency, the pre-Lisbon numeration is used in the discussion below. However, the main dynamics and rationales of the interinstitutional relations are preserved in the Lisbon Treaty. At the end of this section, I will highlight the main innovations introduced by the Treaty of Lisbon.

3.1. Procedural legal basis: article 300 TEC

The internal decision-making procedure for the external relations of the EC in the pre-Lisbon era was described in article 300 TEC. When international negotiations are announced, article 300 TEC only applies when two conditions are fulfilled. On the one hand, the issues dealt with in the international negotiations need to be covered by EC competences. On the other hand, the purpose of the international negotiations needs to be the establishment of an internationally legally binding agreement. As a result, only for the negotiations leading to a treaty, article 300 TEC is the appropriate procedural legal basis. The EC decision-making process for other international negotiations, such as Conferences of the Parties (COPs) or Meetings of the Parties (MOPs), which result in politically binding decisions, is not covered by article 300 TEC.

The rationale behind the provisions of article 300 TEC is the balance of power between the European Commission and the Council of Ministers (Lenaerts, Van Nuffel, Bray, 2005). Article 300 TEC forces the Commission and the Council into an interdependent relation with each other. They need to cooperate if they want to make progress in the EC's external relations and in the EC's presence in multilateral negotiations. I will describe the procedural legal basis of article 300 TEC in a chronological way: first the stage before the international negotiations, then during the international negotiations, and finally after the international negotiations.[5]

The Commission has the exclusive right of initiative under article 300. Hence, the decision-making process starts with an initiative of the Commission, proposing to the Council an authorization to negotiate internationally. However, the right of initiative is *only legally* attributed to the Commission in an exclusive way (Thieme, 2001; Pollack, 2003a; Eeckhout, 2004). As I will elaborate in Chapter 3, the member states or developments at the international level can *de facto* be the starting point of the process.[6]

The Commission's proposal to the Council generally consists of two parts (Macleod, Hendry, Hyett, 1998; Verwey, 2004). On the one hand, the Commission elaborates the background and the reasons of the initiative and the upcoming international negotiations. On the other hand, the Commission includes a draft Council decision, mentioning the substantive legal basis and a proposal for authorization. Once arrived in the Council, the proposal passes through its funnel system: from the Council Working Group – *in casu* the Working Party

---

5   In Chapter 3, these stages will respectively be labelled 'authorization stage', 'negotiation stage' and 'ratification stage'.

6   The member states even have a legal instrument to push the Commission to take legislative initiatives, as article 208 TEC (or current article 241 TFEU) states that '[t]he Council [...] may request the Commission to [...] submit to it any appropriate proposals'.

on International Environmental Issues (WPIEI)[7] – to Coreper and finally to the ministerial level.

The Council then decides on the authorization of the Commission as the EC negotiator. Hence, the Commission negotiates on behalf of the EC for issues covered by Community competences. Besides the authorization decision, the Council can also opt to issue negotiation directives (often referred to as the 'mandate'), which include negotiation instructions for the Commission.[8] The mandate can be considered as a result commitment: it determines the result the Commission has to obtain, but it leaves open the strategy (Thieme, 2001).

The decision-making rule in the Council on both the authorization and the mandate follows the voting rule for similar issues in internal decision-making (Thieme, 2001; Bretherton, Vogler, 2003; Eeckhout, 2004). Hence, when an internal policy measure on a particular issue is decided on by qualified majority voting (QMV)/unanimity, the decision on the authorization and possible mandate is also taken by QMV/unanimity.

Following article 300 TEC, the Council thus appoints the Commission as the EC negotiator. This is the case for international negotiations dealing with issues that exclusively fall under EC competences, or for the issues falling under the EC part of mixed negotiations. Article 300 TEC also stipulates that the Commission negotiates at the international level in consultation with a special committee, which is appointed by the member states. In the field of external trade policy, this committee is called 'Committee 133'[9] (Johnson, 1998; Kerremans, 2004a). Although its composition is not explicitly regulated by article 300 TEC, the committee consists of member state representatives. From that perspective, it can be seen as an emanation of the Council.

Once the negotiators have reached an international agreement, this text is initialled. Initialling, meaning that the negotiators confirm that this text is the result of the negotiations, is a task for the negotiator. Hence, the Commission initials the international agreement on behalf of the EC (Eeckhout, 2004; Verwey, 2004; Lenaerts, Van Nuffel, Bray, 2005). At this point, the international agreement is still a political (legally non-binding) agreement. The next stage from an international law point of view is the signing of the agreement. Signing an agreement means that the parties agree with the final text, which they will submit for ratification at their internal levels. While the initialling of an MEA takes place at the end of the final negotiation session, the signature usually occurs during a formal signing ceremony somewhat later. Article 300 TEC stipulates that the signing of an agreement is the

---

7   'WPIEI' is the generic term for all Council Working Groups dealing with international environmental issues. Depending on the issue discussed, one speaks about WPIEI Climate Change, WPIEI Biosafety, WPIEI Biodiversity, etc.

8   The mandate is not a public document, as it is confidential and thus not published in the Official Journal.

9   This name is derived from article 133 TEC, regulating the decision-making process for trade policies.

prerogative of the Council, on the proposal of the Commission (Leefmans, 1998; Eeckhout, 2004; Verwey, 2004). In practice, this means that the Council designates the Presidency to sign the international agreement on behalf of the EC.

The final stage is the ratification of the international agreement. Ratification has a double significance (Macleod, Hendry, Hyett, 1998; Eeckhout, 2004; Verwey, 2004). On the one hand, it serves as the internal consent to be bound: the EC commits itself to implement the international agreement into the EC legislation. On the other hand, ratification binds the EC vis-à-vis the external parties. The EC ratification follows a similar procedure like the authorization and the signatory decision: the Commission proposes the decision and the Council ultimately decides on it. The decision-making rule (QMV or unanimity) for ratification reflects the rule for internal decision-making and is thus the same like for the authorization. The ratification usually takes the legislative form of a decision that is published in the Official Journal.[10]

Although the new procedural article in the Lisbon Treaty (article 218 TFEU) maintains the main dynamics from the old article 300 TEC, some new elements and modifications have popped up in the new procedural arrangements under the Lisbon regime. First, as I already mentioned, international legal personality and competences have moved from the EC to the EU, as a result of which article 218 TFEU is now about the EU negotiating the issues touched upon by EU competences, instead of the EC negotiating the issues falling under EC competences. Second, article 218 TFEU also foresees to appoint a 'Union negotiator' or 'the head of the Union's negotiating team', which are not further specified in the TFEU. The institutional parameters determining the relation between the EU negotiator and the Council (e.g. authorization on the basis of the negotiator's recommendation, the negotiator's role on signing and ratifying the agreement) are the same for the all possible negotiators. Third, since the Treaty of Lisbon foresees an external representation function for the High Representative of the Union for Foreign Affairs and Security Policy and the European External Action Service, as well as for the permanent President of the European Council (at his/her level), it remains to be seen what their roles will be in the EU's external environmental relations. Now that more and more environmental issues become increasingly politicized and globally handled at the highest political level (think of climate change), future practice will show whether and to what extent these new actors will become involved in negotiating MEAs. Fourth, the Lisbon Treaty changes the ratification procedure for international agreements in the EU. Whereas the Treaty of Nice only stipulates that the Council has to conclude international agreements on behalf of the EC, the Lisbon Treaty adds a veto player in this stage: the European Parliament (EP). Indeed, article 218§6 TFEU prescribes that the consent of the EP is required for concluding 'agreements covering fields to which either the ordinary legislative procedure applies, or the special legislative procedure where consent

---

10   As the international agreement is always published in the Official Journal as an appendix to this decision, it becomes part of the internal legal order of the EC.

by the European Parliament is required'. Future MEAs seem to fall under this category. Before the Lisbon Treaty entered into force – and thus in the period studied in this book – the EP was only involved in the ratification through the consultation procedure.[11] As the consultation procedure implies that the EP can give its non-binding opinion on the Commission's proposal before the Council decides, its formal role was rather limited. Indeed, the EP had the opportunity to give its opinion of the ratification of the EC, but this was not obligatory. After a particular period, the Council could proceed with the ratification even without the EP's opinion (Pocar, 2002; Verwey, 2004).

Although the TEC did not stipulate an important role for the EP before the entry into force of the Lisbon Treaty, an interinstitutional agreement between the Council, the Commission and the EP was agreed upon on a number of procedures. As interinstitutional agreements are located on the border between law and politics, their legal effects and enforcement are rather unclear (Maurer, Kietz, Völkel, 2005). The interinstitutional agreements on this topic that were in force in the period studied in this book were the *Code of Conduct between the Commission and the Parliament* (1995) and the *Framework Agreement between Commission and Parliament* (2000, revised in 2005),[12] which foresees that the EP has to be informed by the Council before the negotiations and by the Commission during and after the negotiations, allowing the EP to hold a debate on the negotiations. Moreover, following this interinstitutional agreement, Members of the European Parliament (MEPs) can be part of the EU negotiation team as observers, without being allowed to take the floor in the international negotiations (Pocar, 2002; Eeckhout, 2004; Koutrakos, 2006).

3.2. Substantive legal basis: articles 174 and 175 TEC

As mentioned above, article 300 TEC/218 TFEU only specifies the decision-making procedure for the EC's/EU's external relations with regard to first pillar issues. It does not provide the permission to act in a particular policy area or with regard to specific issues. Therefore, the EC/EU can only conclude international agreements on the basis of article 300 TEC/218 TFEU *juncto* a second Treaty article providing the substantive legal basis (Macleod, Hendry, Hyett, 1998; Lenaerts, Van Nuffel, Bray, 2005).

---

11   However, parliamentary assent was also needed for the ratification of association agreements, agreements establishing a specific institutional framework by organizing co-operation procedures, agreements having important budgetary implications, and agreements amending EU regulations that have been adopted by codecision.

12   The 2005 Framework Agreement was not the first interinstitutional agreement with regard to the conduct of international negotiations by the EU, since it was preceded by the *Luns procedure* (1964), the *Luns-Westerterp procedure* (1973) and the *Commission's Communication on the Role of the Parliament in the Preparation and Conclusion of International Agreements and Accession Treaties* (1982) (Verwey, 2004; Delreux, 2006; Koutrakos, 2006).

The substantive legal bases for external environmental policy were articles 174 and 175 TEC, which are nowadays articles 191 and 192 TFEU. Article 174 TEC provided the objectives and the action principles of the EC environmental policy. Moreover, paragraph 4 stated that '[w]ithin their respective spheres of competence, the Community and the Member States shall cooperate with third countries and with the competent international organizations'. Article 175 TEC described the decision-making procedure for environmental issues. Council decisions on the authorization, signature or ratification were to be based on either article 175§1 TEC (stipulating QMV) or article 175§2 TEC (prescribing unanimity[13]).

4. The member states as external environmental negotiators

4.1. The option to delegate, but the duty to cooperate

As the member state part of the decision-making process with regard to the negotiation of a mixed agreement is not regulated by the Treaties, there is no strict legal basis for joint member state participation in mixed negotiations. Moreover, as the member states are still competent for the issues involved, they have the possibility to negotiate on their own. Theoretically, member state participation in international mixed agreements is not a question of EU decision-making or EU procedures. The main difference with the EC part, for which article 300 TEC or nowadays article 218 TFEU assigns the Commission as the EC negotiator, is the non-obligatory nature of delegation. Indeed, delegation is optional and voluntary in case of the member states part of the negotiation of a mixed agreement. Member states may opt to pool their voices, but there is no EU framework that obliges member states to do so. If they opt to do so, it is rather on the basis of a political rationale than on the basis of a legal procedure.

However, member states cannot act at the international level in a completely unrestricted manner. Article 10 TEC, replaced by article 4§3 TEU under the Lisbon Treaty, stipulates the principles of Community loyalty and the duty of cooperation, meaning that the member states and the EU institutions are committed to cooperate and to do everything necessary to fulfil the obligations, principles and objectives of the Treaties (Macleod, Hendry, Hyett, 1998; Heliskoski, 2001; Dutzler, 2002; Verwey, 2004). The ECJ has repeatedly used article 10 TEC as a source of inspiration to emphasize the duty of cooperation and community cooperation in

---

13   The environmental issues falling under the unanimity rule are provisions primarily of a fiscal nature; measures concerning town and country planning, land use with the exception of waste management and measures of a general nature, and management of water resources; and measures significantly affecting a member state's choice between different energy sources and the general structure of its energy supply. QMV is applied for all other environmental issues.

external affairs.[14] The Court frequently stated that the member states should strive for close association during the negotiation, ratification and implementation of an international agreement. However, legal scholars still disagree on the question whether article 10 TEC and the case law of the ECJ should be considered as only applicable to the EC part of a mixed negotiations or to the totality of negotiated issues (Eeckhout, 2004).

It is important to note that the duty of cooperation does not imply an obligation to take up a common position. It only means that member states have the duty to strive for a common position. Indeed, the Court mentions 'close cooperation', and for instance not 'unity of representation' (Macleod, Hendry, Hyett, 1998; Verwey, 2004). Moreover, the practical consequences of this duty of cooperation are not clear. What do 'close association', 'duty of cooperation' or 'community loyalty' mean in practical terms (Macleod, Hendry, Hyett, 1998)? It is only certain that the duty of cooperation is legally binding in itself, while the way to realize this is not (Dutzler, 2002).

4.2. The duality of the EU negotiation arrangement in negotiations leading to a mixed agreement

As MEAs are mixed agreements, the legal framework suggests a mixed composition of the EU negotiation arrangement in which both the EC/EU and the member states are represented (Bretherton, Vogler, 2000). The former is represented by the Commission, the latter can (but are not obliged to) be represented by common representative, e.g. the Presidency. There are, however, no procedures on the composition or functioning of the EU negotiation arrangement in negotiations leading to a mixed agreement (Heliskoski, 2001). The discrepancy between, on the one hand, the legal (*de iure*) situation of a strict distinction between an EC/EU part and a member states part and, on the other hand, the political reality in which such distinction is *de facto* less clear and in which the MEA is treated as one entity confronts the EU actors with a challenging situation. Indeed, the result of the duty of cooperation, a vague distinction of competences, and the practical circumstances of international negotiations lead to the following conclusion by Verwey: 'Although in theory both parties [the EC and the member states] are free to pursue their own negotiation, this is never the case' (Verwey, 2004, p. 110). The fact that the representation of the EU in international environmental negotiations has repeatedly been ensured by the troika, consisting of EC and member state representatives, is indicative for this. In environmental affairs, the troika is currently composed of the Presidency, the incoming Presidency and the Commission.[15]

---

14   See particularly Ruling 1/78 (IAEA/Euratom Case), Opinion 2/91 (ILO), and Opinion 1/94 (WTO).

15   In the context of CFSP and before the entry into force of the Treaty of Lisbon, the troika usually referred to the rotating Presidency, the High Representative for CFSP and the Commission. Since the Lisbon Treaty entered into force, references are made to the so-called 'new' or 'informal' troika in the CFSP area, consisting of the President of the

However, before 2001, the Commission was not a member of the troika, since it then was made up of the current, previous and incoming Presidencies.

5. Conclusion: EU decision-making with regard to mixed agreements as opposed to exclusive EC agreements

So far, the literature on the EU decision-making process regarding external first pillar issues mostly focussed on exclusive competences. Studies of the internal decision-making processes in the framework of the non-CFSP external relations of the EU, were mostly limited to the area of trade policy (Meunier, 1998; Coleman, Tangermann, 1999; Collinson, 1999; Nicolaïdis, 1999; Meunier, 2000; Pollack, 2003a; Young, 2003; Kerremans, 2004a; De Bièvre, Dür, 2005; Meunier, 2005; Dür, 2006; Kerremans, 2006; Meunier, Nicolaïdis, 2006; Damro, 2007; Dür, Zimmermann, 2007; Elsig, 2007a; Meunier, 2007; Billiet, 2009). Analysing the internal decision-making process with regard negotiations leading to mixed agreements, in contrast, adds some complicating legal factors to the analysis. In other words, the mixed character of MEAs makes the internal EU decision-making process less straightforward than the decision-making process with regard to the negotiation of an exclusive EC/EU agreement. To conclude this chapter, the main differences between the decision-making process with regard to international negotiations leading to exclusive EC/EU agreements and leading to mixed agreements, such as MEAs, are presented in Table 2.1 and discussed below. These differences are based on elements originating from this chapter's previous sections.

- [DIVISION OF COMPETENCES] The main cause of mixed agreements is the non-exclusivity of the competences. When the competences with regard to an issue dealt with in an international agreement are shared between the EC and the member states, that agreement will be a mixed agreement. By contrast, when the competences are exclusively attributed to the EC, the agreement is not mixed. As a result, the whole puzzle of the EU decision-making process with regard to international negotiations starts with the competence question. The division of competences is the decisive factor for the nature of the international agreement and for the kind of EU decision-making process. As the division of competences determines whether the international agreement will be mixed or not, it also determines the various aspects of the EU decision-making process, which I briefly touch upon below and which are theoretically and empirically elaborated in the next chapters: the legal basis, the right of initiative, the nature of delegation, the EU representation, and the ratification level.

European Council, the High Representative of the Union for Foreign Affairs and Security Policy, and the President of the Commission.

- [LEGAL BASIS FOR THE INTERNAL DECISION-MAKING PROCESS] Article 300 TEC/218 TFEU describes the decision-making procedure for negotiating exclusive EC agreements. By contrast, there is no clear procedure for the negotiation of mixed agreements. The situation is quite ambiguous: article 300 TEC/218 TFEU plays a role, but not in an absolute way. Besides the procedure prescribed in this article, the member states have a large degree of political flexibility to negotiate at the international level for the issues covered by their competences.

- [RIGHT OF INITIATIVE] The initiative to negotiate, sign and ratify an exclusive EC agreement is legally attributed to the Commission. When mixed agreements are negotiated, however, the Commission remains the agenda-setter for the EC part and the member states can legally initiate the process for their competences.

- [NATURE OF DELEGATION] Delegation to the Commission for the negotiation of exclusive EC agreements is obligatory. This also true for mixed agreements – at least for the EC part –, but the mixed nature of the negotiations gives the opportunity to the member states to negotiate on their own or to set up an *ad hoc* negotiation arrangement (for instance delegation to the Presidency, to another member state, etc.). In such a case, delegation is optional.

- [REPRESENTATION OF THE EU IN THE INTERNATIONAL NEGOTIATIONS] In international negotiations touching upon exclusive EC competences, the EC is represented by the Commission. Mixed agreements, on the contrary, are not automatically – or at least not completely – negotiated by the Commission. There are no European rules on how the EU representation in negotiations leading to mixed agreements should look like (Jupille, Caporaso, 1998; Heliskoski, 2001). From a legal point of view, one would expect that the Commission negotiates the issues falling under its competences and that the member states are free to chose their negotiation strategy: negotiating separately or pooling their voices. When they opt for pooling their voices, this results in the (political) delegation of negotiation authority to a common negotiator. The complicating factor of mixed negotiations for the analysis of the internal decision-making process is the presence of member states (or their common representatives) around the external negotiation table. Before the relation between member states and EU negotiator as well as the latter's discretion can be assessed, one first has to know which actor takes the role of EU negotiator. The term 'dual representation' stresses the unique combination of a supranational and an intergovernmental element in the delegation (Sbragia, 1998). Consequently, the clear separation between the actors in charge of negotiation (Commission) and these in charge of authorization and ratification (member states in the Council), like it exists for exclusive EC agreements, disappears in mixed negotiations and the relations between these two types of actors may become more blurred.

- [RATIFICATION LEVEL] Agreements on issues with exclusive EC competences are only ratified by the EC, more in particular by a decision among the member states in the Council (complemented with the EP's consent since the entry into force of the Lisbon Treaty). By contrast, mixed agreements require a double ratification procedure: on the one hand by the EU following its own procedures (i.e. by the collectivity of the member states in the Council in the period analysed in this book) and on the other hand by each member state separately, according to its own constitutional requirements.[16]

Table 2.1   EU decision-making processes with regard to exclusive EC agreements versus mixed agreements

	Exclusive EC agreement	Mixed agreement
division of competences	exclusive EC competences	shared competences
legal basis internal decision-making process	article 300 TEC	partly article 300 TEC, partly unknown: totality unclear
right of initiative	Commission	Commission + member states
nature of delegation	obligatory	partly obligatory, partly optional
EU representation	unity of representation: Commission	dual representation: Commission + member states
ratification level	EU level	EU level + national levels

The legal framework of the EU decision-making process with regard to international negotiations leading to an MEA serves as the starting point to describe the process from a theoretical point of view. In the next chapter, the decision-making process of article 300 TEC is conceptualized in a principal–agent model, to which I also add the complicating factors originating from the mixed nature of MEAs. In Chapters 6 and 7, the legal framework, as elaborated in this chapter, is evaluated. More in particular, for each stage of the decision-making process, I examine to what extent the legal provisions are followed and how the member states and the EU negotiator flesh out their room of political flexibility in practice. The main driving question is how EU actors cope with the tension between the tight legal framework, which suggests a divided decision-making process and negotiation arrangement, on the one hand, and political pressure to negotiate with a high degree of bargaining power, on the other hand.

---

16   The distinction between the collectivity of member states and the member states as individual actors is important because side-payments and package-deals may play a role during ratifications in the Council (the collectivity of member states) and not during ratifications on the different domestic levels.

Chapter 3
Member States as Principals, EU Negotiators as Agents

1. A principal–agent perspective on delegation and discretion

The EU decision-making process with regard to international environmental negotiations can be conceived as a principal–agent relation between the member states, acting as principals, and the EU negotiator, acting as agent. Principal–agent models help to understand situations in which one (set of) actor(s) is acting on behalf of another (set of) actor(s). This is exactly what is happening in the EU when it participates in international negotiations: an EU negotiator is acting, or more in particular 'negotiating', on behalf of the member states. Indeed, member states delegate negotiation authority to an EU negotiator, as a result of which the latter enjoys a certain degree of discretion vis-à-vis the former. Principal–agent models offer concepts and insights, which are relevant to frame relations between actors who delegate (member states, principals) and actors who represent them (EU negotiators, agents). Also the result of delegation, i.e. the degree of discretion enjoyed by the agent, can be analysed by applying a principal–agent framework.

To build a fully applicable principal–agent model that frames the relations between the member states and the EU negotiator in the context of international environmental negotiations, the basic delegation model needs to be extended in three ways. As a starting point, Section 2 discusses the basic model. In Section 3, three theoretical extensions are developed. First, the model is extended by stressing the information asymmetry in favour of the principals and the compellingness of the external environment. Second, the basic principal–agent model, as it has been used to explain EU decision-making on exclusive EC competences (mostly for trade negotiations), is extended to the application on negotiations leading to mixed agreements (and *a fortiori* MEAs). In such negotiations, delegation to the Presidency or to another member state, as well as the domestic ratification requirement come into the picture. Third, the basic model is enriched with insights from the sociological institutionalism (SI), since I do not *a priori* consider the EU decision-making as conflictual, as is assumed by rational choice institutionalism (RCI), from which principal–agent theory is derived. After the development of this enriched principal–agent framework, Section 4 outlines the research design, emphasizing the selection of variables on the one hand, and the case selection on the other hand.

2. The basic principal–agent model, applied to the EU as international negotiator

2.1. Delegation and control

The central concepts in principal–agent models are delegation and control. The basic argument is (1) that principals decide to delegate negotiation authority to an agent; (2) that this generates costs for the principals; and (3) that the principals create control mechanisms to mitigate these costs.

First, member states, as principals, decide to delegate negotiation authority to the Commission (agent) for issues with EC competences. This delegation rationale is laid down in article 300 TEC, nowadays article 218 TFEU, and has been explained in the previous chapter. Moreover, as will become clear in the models' second extension, delegation to the Presidency or a lead country can take place as well. The traditional principal–agent argument is that delegation takes place because it reduces the transaction costs of decision-making. Indeed, principals expect that the benefits of delegation will compensate its costs (Sandholtz, 1996; Epstein, O'Halloran, 1999; Pollack, 2003a). The main benefit in the context of the EU negotiating internationally is the creation of bargaining power by pooling the preferences, voices and votes of the various EU actors (Frieden, 2004). This is the rational choice inspired starting point of principal–agent theory: it assumes a functional logic (Tallberg, 2002). This starting point will be qualified in the third extension of the model.

Second, principal–agent theory assumes that delegation not only generates functional benefits for the principals, but also costs. Generally, it presumes two costs of delegation (Waterman, Meier, 1998). On the one hand, the agent may try to realize his own preferences, which are possibly diverging from the principals'. As a result, delegation can result in opportunistic behaviour by the agent. This cost is called '(agency) slack' or 'agency loss' (Pollack, 2003a; Shapiro, 2005; Hawkins, Jacoby, 2006; Hawkins, Lake, Nielson, Tierney, 2006). On the other hand, the agent has private information, which the principals do not have, in particular about his negotiation behaviour at the international level and the international negotiation process in general.

Third, principal–agent theory claims that principals create control mechanisms to mitigate the costs of delegation and to ensure that the agent performs as loyally and as conformably to the principals' preferences as possible (Waterman, Meier, 1998). The principals thus have to find a balance between sufficient discretion for the agent to assure an optimal bargaining power on the one hand, and not too much discretion to limit the preference and information advantages of the agent on the other hand (Tallberg, 2002). Hence, it is clear that in principal–agent models, delegation and control are inextricably bound up with each other (Christensen, Lægreid, 2007). Together with the agent's behaviour, the combination of delegation and control determines the agent's discretion.

2.2. Two levels, three stages and six control mechanisms

In this section, I apply the principal–agent model to the decision-making process, prescribed by article 300 TEC, where the Commission negotiates on behalf of the member states for issues on which the EC has exclusive competences. In this application, I distinguish two levels, three stages and six control mechanisms (Delreux, 2008).

2.2.1. Two levels

The two levels at which this decision-making process occurs, are the international level and the EU level. At the international level, the international agreement is negotiated. At the EU level, member states decide on delegation and control, and they pool their preferences. The EU negotiator, i.e. the agent, operates at both levels. Two-level-game scholars portray the agent's role as 'Janus-like' (Putnam, 1988; Moravcsik, 1993), as his strategy depends on both the international negotiation context, where his aim is to conduct an international agreement with the external partners, and on the expected behaviour of the principals at the EU level, from which the agent ultimately needs the ratification to fulfil his international commitments. As a result, the EU negotiator is exposed to incentives and constrains from both levels (Milner, 1997). The agent thus has to find a balance between the two levels. The decisive factor in this balance is his discretion, which determines the room of manoeuvre to play both levels against each other (Mo, 1995; Collinson, 1999; Kerremans, 2004b).

2.2.2. Three stages and six control mechanisms

The decision-making process can be analytically divided into three stages: the authorization (t_1-t_2), negotiation (t_2-t_3) and ratification stage (t_3-t_4) (see Figure 3.1) (Nicolaïdis, 1999; Meunier, 2000; Meunier, Nicolaïdis, 2000; Meunier, 2005).[1] In every stage, two control mechanisms are at the principals' disposal: respectively two *ex ante*, two *ad locum* and two *ex post* control mechanisms. The authorization stage starts when the Commission considers a proposal for an authorization decision and possibly a mandate to the Council. However, the Commission has only formal agenda-setting power (Schmidt, 2000; Pollack, 2003a). Informally, the principals can exert pressure on the Commission to initiate the process. Moreover, the Commission's agenda-setting power is particularly undermined by pressure coming from the international level. When an international negotiation session is announced, mostly by the UN or another international organization (under whose auspices the negotiations will take place), the Commission can *de facto* not refuse to initiate the process, if it does not want to remain absent during the negotiations. In such a case, the Commission's right of initiative is prompted and driven by international developments.

---

1   At t_1, the Commission considers a proposal for the delegation decision; at t_2, the first session of the international negotiations start; at t_3, the last international negotiation session ends; and at t_4, the international agreement is ratified by the EC.

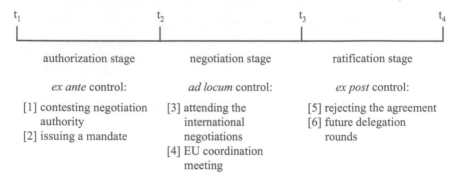

Figure 3.1   Three stages and six control mechanisms in the EU decision-making process with regard to international negotiations

In the authorization stage, the member states build two *ex ante* control mechanisms:

- [CONTROL MECHANISM 1] on the one hand, the member states can *contest the negotiation authority* of the Commission and they are even able not to grant this authority to the agent. When the Council does not authorize the Commission, the process ends here. Non-delegation can thus be considered as the ultimate control mechanism. This is the most powerful tool the principals can use: it reduces the agent's autonomy to nil, as the agent has no authorization to participate in the international negotiations. However, the cost for the member states to use this control mechanism is extremely large as well: they are not represented during the negotiation stage;
- [CONTROL MECHANISM 2] on the other hand, the member states can issue a *mandate* ('negotiation directives') in which they lay down both substantive and procedural instructions for the Commission. Substantive instructions deal with the content of the future international agreement. It gives the Commission an indication of the range of possible international agreements, which will fall within the win-set of a necessary majority of the member states. The procedural instructions in the mandate stipulate the rules of the game the Commission has to respect in his relation with the principals during the negotiation stage.

In the negotiation stage, the Commission negotiates at the international level with the external partners and the member states can use two *ad locum* control mechanisms (Kerremans, 2006; Delreux, 2008):

- [CONTROL MECHANISM 3] member states can sometimes *attend the international negotiations*. This allows them to observe – and control – the agent's negotiation behaviour directly;

- [CONTROL MECHANISM 4] a 'special committee appointed by the Council' has to be consulted by the Commission during the negotiation stage. Although the precise composition and procedures of this committee are not specified in article 300 TEC/218 TFEU, the committee in practice exists of member state representatives (Johnson, 1998). In this committee, which I call the '*EU coordination meeting*',[2] the representatives of the member states meet during the negotiation stage.

In the ratification stage, the internationally negotiated agreement has to be ratified by the member states. This means that the agent has to obtain an approval by the principals on the agreement negotiated with the external partners and on the commitments made at the international level on behalf of the principals. Without this ratification, the EC/EU is not legally bound to the agreement. The ratification stage follows a 'take it or leave it' logic. The member states in the Council can only accept or reject the agreement: they are not able to amend it. In this third stage, member states have a fifth and a sixth *ex post* control mechanism at their disposal:

- [CONTROL MECHANISM 5] on the one hand, the principals can *reject the agreement*.[3] In such situation, the agent faces a so-called 'involuntary defection' (Putnam, 1988; Moravcsik, 1993) as it is not able to fulfil the commitments made vis-à-vis the external negotiation partners. Strictly speaking, involuntary defection means non-ratification in its formal sense, i.e. a failure of the ratification decision by the Council after the agreement is signed. However, in my model, I also consider a non-acceptance of the political agreement at the end of the last negotiation session as an involuntary defection. Key element of an involuntary defection is that an agent is called back by his principals at the moment the agent has made a commitment vis-à-vis the external negotiation partners;
- [CONTROL MECHANISM 6] on the other hand, in the ratification stage, principals can develop their preferences regarding future delegation (Menon, 2003; Kerremans, 2004a). International environmental governance requires regular international negotiations. Principals can take into account their experience with their agent when the delegation question pops up for negotiations on other MEAs. In other words, the situation of iterated (Kerremans, 2004a) or repeated delegation (Bendor, Glazer, Hammond,

---

2   Only in EU trade policy-making, this committee, as it is referred to in article 300 TEC/218 TFEU, is called a 'committee' (more in particular: 'committee 133'). However, in other former first pillar issue areas, the concept 'committee' is not used, but decision-makers call it the 'EU coordination meeting'.

3   Rasmussen indicates that a rejection does not necessarily mean that principals reject the negotiation behaviour of the agent (Rasmussen, 2005). There can be other reasons for a rejection by the member states in the ratification stage. Hence, non-ratification cannot automatically be considered as a sanctioning of the agent by the principals.

2001) or a repeated game (Bueno de Mesquita, 2004) creates a sixth control mechanism.

A key element to understand the EU decision-making process is the anticipation of agents and principals on the future stages. The member states and the Commission base their behaviour on their expectation of what is going to happen in a next stage. This backward inductive reasoning determines the relation between principals and agents. In this sense, I assume that the negotiation behaviour of an agent in the negotiation stage will always be based on his assessment of the chance to get the international agreement ratified in the ratification stage (Milner, 1997; Cutcher-Gershenfeld, Watkins, 1999; Meunier, 2000; Hug, König, 2002; Kerremans, 2004b; Delreux, 2008). Also principals anticipate on the following stages, e.g. when they design the mandate control mechanism in the authorization stage, as they know that they still keep other control mechanisms in reserve. Figure 3.2 brings together the two levels, three stages and six control mechanisms of the article 300 TEC decision-making process, modelled into principal–agent theory.

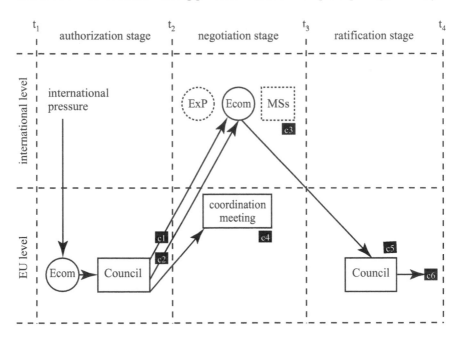

Figure 3.2   Principal–agent model and exclusive EC competences

Note: 'Ecom' = European Commission; 'ExP' = external negotiation partner; 'MSs' = member states; 'c' = control mechanism

3. Extending the principal–agent model

3.1. Extension 1: private information for the principals, cost of no agreement and external compellingness

3.1.1. Preferences, fallback positions and asymmetric information: the perspective of a single member state

I make a distinction between the interests and the preferences of a member state. I follow Milner's definition that interests are the fundamental goals of an actor, which change little (Milner, 1997). Preferences, on the contrary, derive from interests and are the institutional translation of interests (Druckman, Lupia, 2000). Briefly, preferences are modifiable, and are used as positions in decision-making processes in function of both the actor's interest and the (institutional) decision-making context.

I apply an assumption that is often used in political science: actors have single-peaked preferences. This implies that preferences can be represented as Gauss curves (Bueno de Mesquita, 1994; Rasmussen, 2000; Arregui, Stokman, Thomson, 2004). Single-peaked preferences combine the benefits of two conceptions of preferences. First, preferences as *ideal points* (Milner, 1997; Ballmann, Epstein, O'Halloran, 2002; Pajala, Widgrén, 2004) allow the researcher to conceptualize preference distances and to indicate that an actor has a single preference that maximizes his interest. Second, preferences as *win-sets* (Putnam, 1988; Meunier, 2000) allow the researcher to refer to overlapping preferences. Moreover, this conception of preferences explains that an actor will not only accept an outcome that exactly corresponds to its interest, but also outcomes that are positioned closely to these interests but (slightly) less desired.

Single-peaked preferences of a member state are presented in Figure 3.3. On the X-axis the different policy options are situated, going from the status quo (left) over minimalist to more maximalist policy options (right). The Y-axis displays the degree of preference intensity that a member state attributes to a particular policy option. This is the extent to which a member state is prepared to invest a particular cost in defending the corresponding policy option. In other words: the higher the intensity, the more a member state wants a particular policy option. As a result, the higher a point of the Gauss curve is situated above the X-axis (i.e. if preference intensity > 0), the more a member state supports that policy option. The lower a point of the Gauss curve is situated below the X-axis (i.e. if preference intensity < 0), the more a member state opposes that policy option.

The projection of the top of the Gauss curve on the X-axis is the ideal point of that member state. Every outcome under the curve (the win-set, i.e. the projection of the curve on the X-axis) is still acceptable for that member state. The win-set is composed of the various preferences, which have an intensity (a value) that decreases as the distance to the ideal point increases (Jupille, 1999; Arregui, Stokman, Thomson, 2004). Consequently, in this conception of preferences, preferences have three dimensions: ideal points, spread and intensity.

The frontiers of the win-set are formed by the intersection of the Gauss curve and the X-axis. These are the policy options with a preference intensity of zero, regarding to which the member states are indifferent.

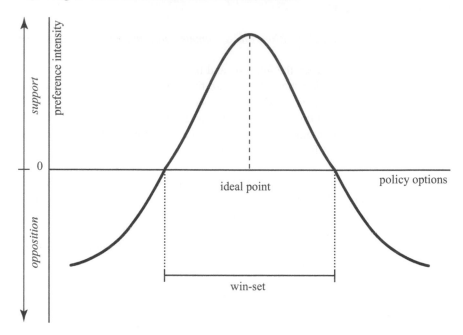

Figure 3.3   Single-peaked preferences as combinations of ideal points and win-sets

The situation presented in Figure 3.3 only applies in the *authorization stage.* In this stage, the principals make their preferences clear to each other and to the agent. Their goal is to guide their agent by indicating which outcomes they will be able to accept in the ratification stage. However, a principal does not only have preferences, but also fallback positions. This creates a situation of asymmetrically divided information in favour of the principals. Indeed, the principal–agent model is not only characterized by an information benefit for the agent, but also by an information benefit for the principals (Maskin, Tirole, 1990; Meunier, Nicolaïdis, 2000; Delreux, 2008). While agents are better informed about the substance and the developments of the international negotiations, principals are better informed about their final win-set, i.e. about the range of agreements they are ultimately able to ratify. I assume that member states have some additional room outside their Gauss curve to accept an agreement that is positioned there, when it really comes to the crunch in the ratification stage.

Why would member states ultimately ratify an agreement outside their initial win-set and approve an agreement that is not covered by their Gauss curve? The answer is that such an involuntary defection generates a political cost of no agreement (Moravcsik, 1993; Collinson, 1999). Besides the substantive costs and benefits of not ratifying an MEA, the political pressure not to jeopardize the negotiation process is enormous in a multilateral negotiation context. This pressure creates the cost of no agreement. Not only the time and (financial) resources that are spent in the negotiation process, but particularly the political pressure from the international level not to be held responsible for a failure of the process generates a 'compelling external environment' (Kerremans, 2003; Delreux, 2008). Indeed, a status-quo default condition *de facto* hinders the ultimate sanctioning of the agent (Pollack, 2003a). The compellingness of the external environment depends on the number of third parties involved in the negotiations and their relative bargaining power (the more parties involved and the larger their relative bargaining power, the larger the degree of compellingness).

Briefly, the reasoning I add to the basic principal–agent model goes as follows: a compelling external environment increases the political cost of no agreement for a principal. This leads to principals revealing their private information on their fallback positions, which broadens their final win-set. Consequently, I argue that the range of possibly ratifiable agreements in the ratification stage is larger than what seemed to be acceptable in the authorization stage.

This means that in the ratification stage, the Gauss curves of the member states cover a broader range of policy options than the member states had indicated before. The parts that supplement the initial win-set are the fallback positions. Consequently, in the ratification stage, the compelling external environment forces the member states to put all their cards on the table. I present this graphically in Figure 3.4: the axis on which the Gauss curve is projected moves down, i.e. to a negative preference intensity. This means that the outcome of the international negotiations is not evaluated anymore only against the policy options to which a member state has a positive preference intensity. As a result, the Gauss curve expands and policy options with a certain negative value on the preference intensity will be ratified as well.[4] The policy options corresponding to such a negative preference intensity are not 'wanted' by a member state, but they are covered by the fallback positions of the member states. The intersections of the Gauss curve and the lowered X-axis are located further away from the ideal point than the initial intersections (i.e. when the Gauss curve was projected on the axis that had the value 0 on the degree of preference intensity), resulting in a larger win-set. The extent to which the axis on which the Gauss curve is projected moves down is a function of the cost of no agreement: the higher this cost, the more this axis moves down, and thus the more outcomes that will be ratifiable.

To illustrate these dynamics, I consider three possible outcomes of the international negotiation process: A, B and C (see Figure 3.4). If the international negotiations result in agreement A, a member state assesses this with a positive

---

4   This only holds for policy options covered by the fallback position. Hence, only points with a certain degree of negative preference intensity, depending on the cost of no agreement, will be covered by the ratification win-set.

preference intensity (a'), so this agreement is supported by that member state and it will be ratified without any problem, since it is covered by the win-set during the ratification stage (and also by the win-set of the authorization stage). Agreement B confronts a member state with a particular cost, as its preference intensity is lower than zero (b'). However, the cost of rejecting B (the cost of no agreement) is larger than the cost related to the negative preference intensity. In this situation, the agreement is not covered by the member state's initial win-set, but it is part of the fallback position, as a result of which it also belongs to the win-set used in the ratification stage. Hence, agreement B will ultimately be ratified. Finally, agreement C also corresponds to a negative preference intensity (c'), but in this case, the cost of no agreement is insufficiently large to counterbalance it. As a result, the cost connected to a rejection of the agreement is smaller than the cost connected to the substance of the agreement. C is not covered by the ratification win-set and will thus not be ratified.

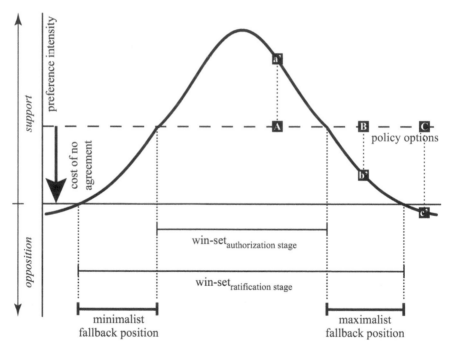

Figure 3.4   Effect of the compellingness of the external environment and the political cost of no agreement on the member states' preferences in the ratification stage

3.1.2. Aggregation of preferences: the perspective of the Council as collective principal
In the previous section, I described the decision-making process from the perspective of a single member state. However, as the EU negotiator represents

the aggregation of the member states, the Council acts as a collective principal (Nielson, Tierney, 2003; Lyne, Nielson, Tierney, 2006; Elsig, 2007a), consisting of several member states, each with their own preferences, Gauss curves and fallback positions. How do these different member state preferences aggregate? For reasons of my model's comprehensibility, I limit the number of member states in the Council to three in this theoretical model. However, my model also holds for a Council with 12, 15, 25, 27 or more member states. As I argued in Chapter 2, there are two possible aggregation mechanisms in the EU decision-making process with regard to international negotiations leading to an MEA: unanimity (article 175§2 TEC/192§2 TFEU) or QMV (article 175§1 TEC/192§1 TFEU). In my model, I conceptualize unanimity as 'all three member states have to approve the agreement', and QMV as 'two of the three member states have to vote in favour'. The aggregation of preferences occurs both in the authorization stage and in the ratification stage.

An agreement is reached in the Council when a certain number of Gauss curves overlap (two of the three with QMV, all three with unanimity).[5] When there is no overlapping between the Gauss curves of the required number of member states to reach QMV/unanimity on t_1 (for the authorization) or on t_3 (for the ratification), preference changes are needed to create such overlapping. In this sense, decision-making processes in the Council are thus meant to move the member states Gauss curves closer to each other. The form of a Gauss curve can change by a modified preference intensity (in the direction of the Y-axis), a changed institutional translation of the interest, or a transformation of the spread of the Gauss curve (both in the direction of the X-axis).

Figure 3.5 depicts the projection of each member state's Gauss curve on the X-axis, both in the authorization and in the ratification stage. This figure allows me to examine the effect of two factors on the aggregation of preferences to a collective Council win-set: the stage in the decision-making process and the voting rule. Two conclusions can be derived from the figure. First, keeping the voting-rule constant, the collective win-set is larger in the ratification stage than in the authorization stage. This proves my argument that the Council will ratify a broader range of international agreements than it has suggested in the authorization stage because of the compelling external environment and the cost of no agreement. Second, a more stringent voting-rule (*in casu* unanimity instead of QMV) generates a smaller collective win-set. This reflects findings from other scholars (Milner, 1997; Meunier, 1998; Jupille, 1999; Meunier, 2000; Young, 2003; Franchino, 2007) and is the case in both stages. It is easier to reach agreement in the Council on the authorization of the EU negotiator and on the ratification of the international agreement when QMV is applicable than when unanimity is the voting rule.

---

5 In fact, it is not the Gauss curves that overlap, but the projections of the Gauss curves on the X-axis.

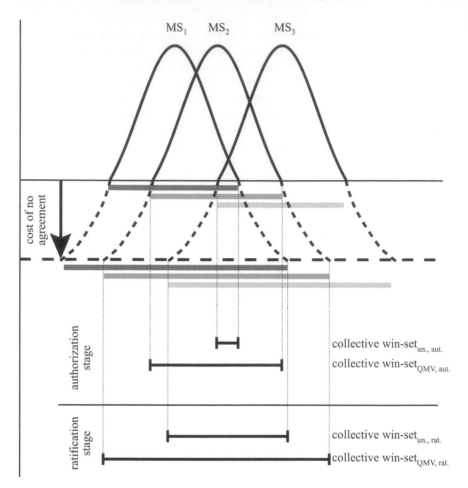

Figure 3.5   Effect of the stage in the decision-making process and the effect of the voting rule on the collective win-set in the Council

3.1.3. The effect on discretion: the perspective of the agent

As mentioned, the EU negotiator can have his own preferences and win-set. This means that the EU negotiator may negotiate at the international level with a double – and often ambiguous – purpose. On the one hand, the EU negotiator aims to reach an agreement with the external partners in accordance with his own preferences. On the other hand, the EU negotiator is concerned about the fact that the international agreement has to fall within the collective ratification win-set of the Council in order to avoid an involuntary defection. This consideration makes the agent's negotiation strategy and representation behaviour very delicate. It is as though the EU negotiator is attached to the Council by a length of elastic. The expected negotiation behaviour of the EU negotiator is to pull the elastic sufficiently tight that the international outcome covers his own win-set, but at the same time not so tight that the elastic breaks and the international agreement is not covered anymore by the Council's ratification win-set.

If the EU negotiator aims to fulfil his agent role successfully, the international agreement has to be ratified by the principals at the EU level. How can the agent increase the likelihood of ratification? The key is transmission: the EU negotiator's strategy is then to optimize the transmission of the external compellingness from the international level to the EU level. By doing so, the agent maximizes the political cost of no agreement for the principals, as a result of which the final collective win-set of the principals expands. This means that the international negotiation context, which is experienced by the EU negotiator, has to be transmitted to the member states. The EU negotiator has to persuade the member states that the reached agreement is the best possible option, i.e. the outcome that maximizes the member states' preferences, given the opportunities and constraints generated by the international negotiation context. This way, by anticipating on the private information of the principals in the ratification stage, the EU negotiator is able to increase his discretion (see Figure 3.6).

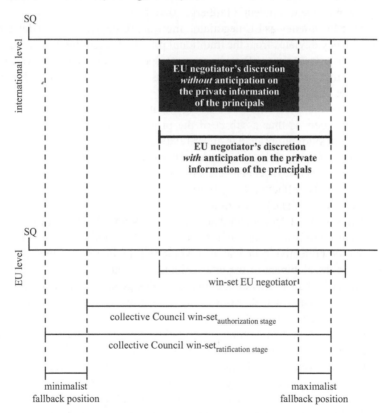

Figure 3.6　Discretion of the EU negotiator, with and without anticipating on the private information of the principals

3.2. Extension 2: mixed agreements, agents as a subset of the principals, and the domestic ratification requirement

As mentioned in Chapter 2, MEAs are mixed agreements, as they deal with issues on which the competences are shared between the EC/EU and the member states. This has a double consequence for my model. First, the Commission is not the only possible agent anymore, as also the Presidency or another member state, a so-called 'lead country' (because it takes the lead and negotiates on a particular topic) can be appointed as agent. Second, mixed agreements require ratification at the international as well as the EU level, which creates an additional, third level in the model.

3.2.1. The Presidency or a lead country as agents
I argue that delegation to the Presidency or a lead country does not follow the same logic as delegation to the Commission. From a theoretical point of view, and extending it to delegation to lead countries, I deviate from Tallberg's argument that 'EU governments' engagement of the Presidency as their representative leads to a classic principal–agent problem' (Tallberg, 2006, p. 141).

As for the Presidency as EU negotiator, there are three reasons why delegation to the Presidency deviates from the traditional principal–agent model: the agent is a subset of the principals, the Presidency rotates six-monthly, and the Presidency plays a particular institutional role. The first reason also holds for a lead country taking the role of agent.

First, when the Presidency acts as agent, it still remains a principal. The Presidency as agent is thus a subset of the principals. Principal–agent scholars have just begun studying principal–agent relations in which the agent is also part of the collective principal that grants negotiation authority to that agent (Delreux, 2008; Hodson, 2009). I expect that delegation to a colleague-principal will mitigate an assumed negative effect of delegation. The information about the international negotiation context will be less asymmetrically divided in favour of the agent when the Presidency is the EU negotiator than when the Commission takes this role. This assumption also holds for a lead country acting as EU negotiator. It is plausible to assume that the collective principal as a whole will be better informed about the substance and development of the international negotiations when the agent is a principal than when the agent does not belong to that group of principals. An agent that is a subset of the principals has a stronger incentive to share information with other principals than an agent that is not.

This incentive is linked with the second characteristic of the negotiating role of the Presidency: as the Presidency rotates six-monthly, the likelihood of agency slack reduces as well. A Presidency is the EU negotiator only for six months. Here, the shadow of the future generates a reciprocity-based principal–agent relation (Axelrod, 1981; Elgström, Jönsson, 2000; Tallberg, 2003). This means that the Presidency as agent is not expected to behave as opportunistically as an agent that is not involved in a rotation system would do. As the negotiations of an MEA mostly take multiple years, an MEA will never be negotiated by one

Presidency. Member state A, holding the Presidency during the first sessions, is well aware that the final deal will be negotiated by colleague-member state B, then holding the Presidency. As A will be a principal during the decisive session in the negotiations, it wants to avoid that B will conquer more autonomy at that time. To avoid that B applies a tit-for-tat logic in the future, A will not conquer more autonomy. Consequently, the rotation system of the Presidency makes the traditional slack problem less likely than it appears in a situation of exclusive EC/EU competences and the Commission negotiating. This argument is supported by research on the EU Presidency, arguing that member states holding the Presidency tend to put aside their national preferences during the course of their Presidency (Svensson, 2000; Schout, Vanhoonacker, 2006). In the same sense, it is reasonable to presume that Presidencies share more information with the member states than the Commission would do, as a Presidency knows that it will be in the position of a mere principal in the near future and that it wants to be informed by the then current Presidency. In this reasoning, the Presidency expects reciprocity by the future Presidencies.

Finally, the Presidency that acts as EU negotiator is not only a member state taking the role of agent, it also fulfils a particular institutional role. This institutional component generates expectations by the other member states about the Presidency's negotiation behaviour. Because of the impartiality and neutrality norm related to the Presidency's role, member states expect the Presidency to be a loyal agent (Metcalfe, 1998). The Presidency also has a range of institutional tools at its disposal, which the Commission or a lead country does not have (Fernandez, 2008). By chairing the Council and coordination meetings, tabling proposals, and having the right to determine a consensus or call for a vote, the Presidency as agent generates additional dynamics in comparison with the Commission or a lead country negotiating. Moreover, Presidencies have an incentive to reach an agreement, both internally among the principals and externally with the negotiation partners, during the six-month course of their Presidency, since the (perceived) success of a Presidency is often assessed in terms of bringing decision-making processes to a favourable conclusion.

Not only member states in their capacity as Presidency, but also member states as lead countries can act as EU negotiator. Granting negotiation authority to a lead country generates the same effect as I mentioned above in the case of the Presidency: as a lead country negotiates while being a subset of the principals, an information advantage for the agent will be less likely. Moreover, delegation to a lead country implies an additional element in the authorization stage: the question of *to which* member state negotiation authority will be delegated. Indeed, the decision of delegation to a lead country is theoretically characterized by a pool of possible lead countries. If this is the case, screening and selection procedures become relevant (Kiewiet, McCubbins, 1991; Hawkins, Jacoby, 2006; Hawkins, Lake, Nielson, Tierney, 2006; Franchino, 2007). This means that the characteristics – both preferences and capabilities – of the potential agents are taken into account by the principals when they decide to authorize. As a result, the first control mechanism (the authorization) is supplemented by a selection aspect: a lead country is not only authorized, but also selected out of a pool of candidates.

3.2.2. The domestic level as additional ratification level

As mixed agreements do not only touch upon EC competences but on national competences as well, they also have to be ratified at the level of each member state. This means that the principal–agent model for negotiations leading to a mixed agreement has to be extended with the domestic level in the third stage. As a result, the two-level game (with the international and the EU level) becomes a three-level game, including the domestic level as well (Pattersson, 1997; Collinson, 1999; Pollack, 2003a; Leal-Arcas, 2004). Figure 3.7 builds upon the principal–agent model for exclusive competences (as presented in Figure 3.2) and extents it to a three-level-game in which the Presidency and/or a lead country also acts as EU negotiator.

I do not consider the involvement of national parliaments as an analytically distinct control mechanism. I classify it with the (fifth) involuntary defection control mechanism, which can thus be deployed at the EU level (in the Council) and at the various domestic levels (in the national parliaments). It is plausible to expect that the addition of a third level makes it harder for the agent to take the ratification hurdle because of two reasons. On the one hand, as the number of veto players increases, the likelihood that the international agreement will be ratified decreases (Tsebelis, 2002). On the other hand, it is very difficult for the EU negotiator to transmit the external compellingness to the domestic level, as he has no direct links with the domestic levels. In this sense, directly transmitting the external compellingness to the national parliaments does not lie within the EU negotiator's reach. By contrast, the agent is more likely to be able to transmit this compellingness to the member states at the EU level. Hence, the involvement of national parliaments in the ratification stage generates a certain degree of uncertainty for the agent, which includes a risk element in his negotiation behaviour.

A result of the inclusion of the domestic level in the principal–agent model is that an EU negotiator, involved in international negotiations leading to a mixed agreement, *de facto* has to anticipate on a collective Council win-set in the ratification stage established by unanimity, even when QMV is the appropriate and formal voting rule in the Council. The reason is that every member state keeps veto power in reserve at its domestic level. In case when article 175§1 TEC/192§1 TFEU is the legal basis of the ratification decision, a member state will not be able to block the EC ratification, but it is able to prevent being bound by a veto at the domestic level.

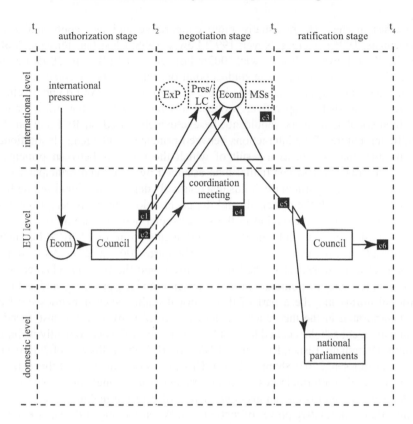

Figure 3.7   Principal–agent model and shared competences

Note: 'Ecom' = European Commission; 'ExP' = external negotiation partner; 'MSs' = member states; 'Pres' = Presidency; 'LC' = lead country; 'c' = control mechanism

3.3. Extension 3: EU decision-making in an institutionally dense environment

3.3.1. Opening the principal–agent model for sociological institutionalist insights
Traditional principal–agent models originate from rational choice institutionalism (RCI) (Worsham, Eisner, Ringquist, 1997; Moser, Schneider, 2000; Aspinwall, Schneider, 2001; Tallberg, 2002; Kassim, Menon, 2003; Pollack, 2003a). It starts from the assumption that actors (principals and agents, member states and EU negotiators) have fixed, exogenous preferences, based on interest maximization and unaffected by the institutional and normative context in which the decision-making process takes place. This leads to a model of decision-making that more readily assumes conflicting preferences and that is characterized by a hard bargaining atmosphere.

However, various studies show that EU decision-making processes cannot be reduced merely to such rational choice inspired assumptions. Decision-making processes in the EU are usually more cooperative and less conflictual (e.g. Kerremans, 1996a; Joerges, Neyer, 1997; Lewis, 1998; Elgström, Jönsson, 2000; Lewis, 2000; Lewis, 2003a; Lewis, 2003b; Pollack, 2003b; Beyers, 2005; Lewis, 2005; Rasmussen, 2005). These studies point out that insights from sociological institutionalism (SI) also play a role in EU decision-making. As I do not exclude *a priori* that such cooperative, problem-solving oriented, dynamics matter, the theoretical framework should not be exclusively based on RCI and on the conflicting nature of the interactions between principals and agents. Institutional norms may matter too, as a result of which the relations between principals and agents may be less conflictual than assumed by traditional principal–agent models. This claim particularly holds in a relatively dense institutional setting like the EU, where both the principals and the agent belong to it. As Shapiro notes, principals and agents belonging to the same 'social network' are more likely to 'share similar interests and values […], and agents are more likely to be other-regarding (altruistic, even) or honest […]' (Shapiro, 2005, p. 277). As I use the theoretical framework, established in the current and the previous chapter, as a tool to understand the political reality of decision-making processes and as these informal norms may be a part of this political reality, such dynamics need to be incorporated in the theoretical model as well. Moreover, in the last decade, principal–agent scholars called to broaden the model with sociologically inspired insights (Worsham, Eisner, Ringquist, 1997; Smyrl, 1998; Tallberg, 2002; Pollack, 2004; Rasmussen, 2005; Shapiro, 2005). This extension aims to meet this call.

SI argues that informal rules and appropriateness are strongly underexposed by RCI. Its main claim is that an accurate picture of decision-making processes has to include the normative power of institutions. When looking at decision-making processes from an SI point of view, decision-makers do not base their preferences on individual interest maximization (like RCI assumes), but on their assessment of the appropriateness of defending a particular preference. Briefly, they search for the most appropriate way of (re)acting, given a particular situation in a particular institutional context. Hence, the decision-making behaviour of principals and agents is based on the norms carried out by the institutional environment.

3.3.2. Institutional norms in the principal–agent relation in the EU

What are exactly these norms in the decision-making context in which principals and agents interact? Mainly following Lewis' studies on decision-making in the Council – and more specifically Coreper – (Lewis, 1998; Lewis, 2000; Lewis, 2003a; Lewis, 2005), I distinguish five institutional norms of the Council: institutional memory and diffuse reciprocity, trust and credibility, mutual responsiveness, consensus and compromise striving, and avoiding process failure.[6]

---

6   In his original work, Lewis distinguishes the following five 'performance norms of the community method': diffuse reciprocity, thick trust, mutual responsiveness, consensus-reflex, and culture of compromise. For the purpose of this study, I combine consensus and

First, institutional memory and diffuse reciprocity are based on the idea that institutions provide a carrier for institutional memory (Martin, Simmons, 1998; Lewis, 2000). This works in two directions. On the one hand, current behaviour of an actor in an institutional context is influenced by the 'shadow of the future' (Elgström, Jönsson, 2000; Tallberg, 2003). The institutional mechanism behind the shadow of the future is 'diffuse reciprocity' (Lewis, 2000), meaning that a decision-maker does not base his behaviour on the expectation of ensuring rewards from other decision-makers, but on a tacit expectation that this behaviour is 'in the interest of continuing satisfactory overall results for the group' (Keohane, 1986, p. 20).

On the other hand, the 'shadow of the past' inspires the EU decision-making process with SI elements (Elgström, Jönsson, 2000): previous decision-making rounds among the same actors within the same institutional setting can influence the current decision-making process. Indeed, common norms and values are likely to develop among decision-makers who cooperate in a densely institutionalized setting. From this perspective, decision-makers act on the basis of appropriate norms, these being developed in the course of the (bygone) process.

Second, trust and credibility can determine the principal–agent relation since the EU decision-making processes with regard to international negotiations leading to an MEA mostly take multiple sessions over several years. Moreover, within the same institutional context of the EU, other (external environmental) policies are negotiated as well. The fact that EU decision-making is repeating – and in a certain sense also continuous – provides that mutual respect, close interpersonal relations and trust are developed (Lewis, 2000).

A situation where A trusts B can be defined as a situation where 'A predicts that B will at least do no harm in a circumstance in which A's interests depend on B's behaviour' (Hoffman, 2002, p. 379). Due to the continuous decision-making setting and the trust among decision-makers, a decision-making context is developed in which the arguments, negotiating behaviour and preferences of the various participants are considered as credible. If principals and agents trust each other, they consider each other's claims as credible, which may affect the principal–agent relation, as well as the agent's discretion.

Third, mutual responsiveness refers to the tacit duty decision-makers may feel about justifying their positions, preferences and strategies vis-à-vis each other. SI argues that member state representatives understand the need to recognize each other's domestic problems and to know each other's national backgrounds in order to reach an agreement. The institutional norm triggering this dynamic is called 'mutual responsiveness' (Lewis, 2000; Lewis, 2003b; Beyers, 2005).

Fourth, decision-making in the Council is characterized by attempts to reach a compromise and by a continuous striving for consensus instead of punctiliously sticking to the formal decision-making rule, although the decision-making rule (QMV or unanimity) in the Council is clearly defined in the legislative act under

---

compromise, I add credibility to the trust norm, and I include avoiding process failure as an additional EU decision-making norm.

consideration (*in casu* the authorization and ratification decision, see Chapter 2) (Kerremans, 1996a; Van Schendelen, 1996; Johnson, 1998; Peterson, Bomberg, 1999; Lewis, 2000; Pollack, 2003b). The rationale behind striving for consensus, even when QMV is the rule, is that member states want to avoid isolating another member state. The decision-makers try to get everyone – or at least as many actors as possible – on board. Consequently, the outcome of a decision-making process in the Council often takes the form of a compromise (Hayes-Renshaw, Wallace, 1997; Arregui, Stokman, Thomson, 2004).

Fifth, avoiding process failure can play a role in the EU decision-making process. As discussed in the first extension of the principal–agent model, the cost of no (international) agreement is theoretically likely to affect the behaviour of the member states in the ratification stage. However, not only 'no *international* agreement' is an important factor in the EU decision-making process, but the member states also aim to avoid the lack of an *internal* agreement in the EU (e.g. on authorization, a common EU position, etc.). This may not only play a role as 'cost of no internal agreement', but avoiding a failure of the decision-making process may also play a role as an institutional norm. Member states may feel a certain responsibility to bring the process in which they are all involved to a successful conclusion. They started the process together and they want to end it together as well. Indeed, the institutional setting and the fact that the decision-makers want to conclude the decision-making process successfully, may create a spirit in which reaching *an* agreement is a driving force, apart from whether this means a political cost or not.

3.3.3. An SI perspective on delegation to an EU negotiator in international negotiations leading to an MEA

In the EU decision-making processes with regard to international (environmental) negotiations, two types of relations can be distinguished: the relations among the member states (principals) mutually, and the relations between the member states as a collectivity (principal) and the EU negotiator (agent). The principal–principal relations – i.e. 'standard' EU decision-making among member states – cannot be fully understood if it is only analysed from an RCI perspective. As the Council is potentially characterized by a dense institutional environment, relations among the member states may have problem-solving, cooperative characteristics, triggered by institutional norms.

As for the principal–agents relation, I follow Rasmussen's critique on the RCI bias of delegation theory: 'there may also be other types of delegation where emphasis is more on the social relation between principals and agents, which may lead agents to be less inclined to maximise their individual policy goals and instead act according to a logic of appropriateness' (Rasmussen, 2005, p. 1029). The rest of this section deals with the different RCI and SI interpretations of the relations between principals and agents in the EU decision-making processes studied. Before describing each stage of the decision-making process from an RCI and from an SI perspective, I present the main differences between RCI and SI for my principal–agent model in Figure 3.8, which elaborates further on the general scheme of delegation in the EU regarding international negotiations leading to an MEA (Figure 3.7).

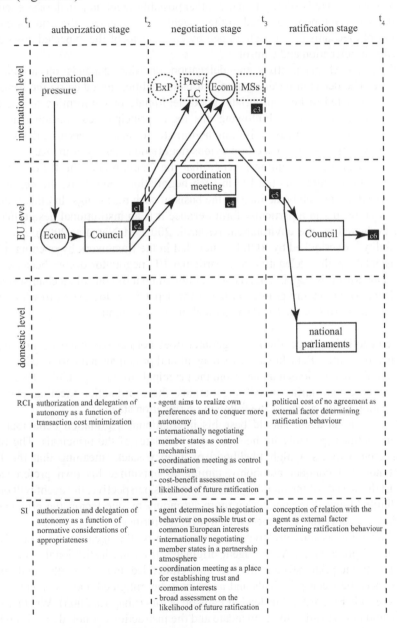

Figure 3.8   Principal–agent model and shared competences from an RCI and an SI perspective

Regarding the authorization stage, RCI argues that autonomy will be granted in such a way the principals' transaction costs are minimized. From this perspective, the decision on delegation depends on the net benefits of delegation to a particular negotiator compared to delegation to another possible negotiator. In the same sense, the degree of autonomy granted to their agent – or, inversely, the establishment of control mechanisms – is determined by an analysis of the expected costs and benefits of delegation and control.

From an SI perspective, the delegation decision depends on normative pressure. The decision on delegation and its modalities are taken on the basis of what is expected to be appropriate, good, legitimate or conforming to a given set of norms and values. Following this logic, principals can reason that e.g. delegation to the Commission is more legitimate, for instance because delegation to the Commission took place in previous negotiation rounds in the same policy area or because it is common practice that the Commission, the Presidency or a lead country represents the EU in such negotiations. Similarly, the degree of granted autonomy can be defined on the basis of such reasonings. In other words, delegation occurs in a particular form because it is an institutional design that is well evaluated by the environment (Gilardi, 2002; McNamara, 2002). Another possibility – overlooked by RCI, but included in the framework by opening it for SI insights – is that delegation to a particular EU negotiator occurs because the principals trust this negotiator. Also the establishment of control mechanisms and the degree of autonomy granted to the agent depends on the extent to which the institutional norms of the EU decision-making process matter.

In the negotiation stage, the EU negotiator does not behave independently from the member states. Four links between agent and principals have to be taken into account: the instructions/mandate from the principals for the agent, direct member state involvement during mixed negotiations, the coordination meeting, and the anticipation by the agent on the following ratification stage.

First, a loyal agent should base his international negotiation behaviour on the instructions (possibly in the form of a mandate) of the principals. The RCI perspective expects a high likelihood of agency slack, meaning that the EU negotiator will conquer autonomy, aiming to maximize his own preferences, against the wishes of the member states. From this perspective, the agent will only feel limited himself in conquering more autonomy by one factor: the principals' readiness to ratify the agreement afterwards (see further).

SI does not assume preference maximization by the agent at the cost of the principals' preferences. As the agent originates from the institutional context in which both the principals and the agent are embedded, the potentially conflicting preferences between principals and agent may be mitigated and the agents may be more 'other regarding' instead of egoistic actors (Shapiro, 2005). Whether the EU negotiator will stick to the mandate and the instructions or not, depends on his relation with the member states. And whether this relation will be conflictual or cooperative depends on the density of the institutional environment principals and agents are operating in. When principals and agents trust each other and when they believe that they are both striving for the same (European) interest or when they conceive themselves as 'acting under one single European identity', agency slack will be less likely (and it will certainly not be driven by interest maximization). In other words, the same normative factors and mechanisms can play a role between member states and the EU negotiator as the institutional norms that influence the relation among the member states mutually. This means that diffuse reciprocity, trust and credibility, mutual responsiveness, consensus and compromise, and avoiding process failure can influence decision-making and interactions between principals and agents in the negotiation stage. Trust between principals and agents is of course very important from an SI perspective on delegation, as it refers to the 'willingness to place the fate of one's interests under the control of others' (Hoffman, 2002).

Second, when mixed agreements are negotiated, one or more member state(s) – more in particular the Presidency, a lead country, or even the member states separately if delegation for national competences does not take place – will negotiate at the international level next to the Commission (see Chapter 2). From an RCI point of view, the presence of (a) member state(s) as negotiator(s) at the international level is considered as a control mechanism, reducing the degree of discretion enjoyed by the Commission (Nicolaïdis, 1999). Indeed, it is more difficult for the Commission to realize its own preferences at the cost of these of the member states with one or more principals on his side.

By contrast, SI emphasizes the possibility that an atmosphere of partnership is established between principals and agent. In such a partnership atmosphere, the normative environment matters as described above: not only between the principals mutually, but also between principals and agents. This means that it is more likely that the agent negotiates on the basis of a common European interest, rather than on the basis of the maximization of his own interests.

Third, the same reasoning holds for the functioning and the role of the EU coordination meeting. RCI considers the coordination meeting as a control mechanism for the principals. Its main task is to monitor the negotiating agent and to avoid agency slack. Again, the potential divergence between the preferences of the member states and those of the EU negotiator and the uncertainty of the member states about the negotiation behaviour of the agent at the international level are the starting points for the analysis of the role and the functioning of the coordination meeting.

The SI point of view, however, starts from a less conflicting assumption of the relations between the member states in the coordination meeting and the EU negotiators. Indeed, SI does not exclude that this relation is cooperative (and for instance based on trust and a common aim of realizing the 'European interest') and that the principals develop a kind of empathy for the situation with which the agent is confronted at the international level. In other words, the coordination meeting functions as a forum for close cooperation between agent and principals, striving for a general EU interest.

Fourth, the agent anticipates on the behaviour of the principals in the next stage (ratification stage). Will the Council ratify the agreement – taking into account both the cost of no agreement and their substantive preferences on the provisions of the MEA – or will the agent face an involuntary defection? From an RCI perspective, the EU negotiator will strictly assess the costs and benefits of his negotiating behaviour at the international level and the expected results for the behaviour of the member states in the ratification stage. In other words, the agent's anticipation will be based on a cost-benefit analysis.

From an SI point of view, the EU negotiator will also make an assessment of the likelihood that the negotiated agreement will be ratified. However, this assessment will be broader than strictly based on an instrumental cost-benefit analysis. This means that a partnership atmosphere, diffuse reciprocity, trust and credibility, mutual responsiveness, the fact that agents and principals may see each other as colleagues striving for consensus and a compromise, and aiming to avoid process failure may also play a role in the agent's anticipation on the ratification stage.

In the ratification stage, the decision-making process moves back from the international level to the EU level. The focus shifts from interactions between member states and EU negotiators towards interactions among the member states mutually within the collective principal. In other words, in the ratification stage, the process is again intra-EU decision-making, like in the authorization stage. However, the presence of the international negotiation context, carrying out a particular degree of compellingness, remains present. The compellingness of the external environment increases the likelihood that an agreement that falls outside the initial win-set of the principals will be ratified. Rational choice institutionalists assess the effect of the compelling external environment – again – in terms of a cost-benefit analysis: when the cost of no agreement increases, their new win-set broadens.

From an SI perspective, the dynamic of the compelling external environment and the cost of no agreement counts as well. However, decision-making in the ratification stage not only depends on its costs and benefits of ratification for the realization or maximization of the principals' preferences, but also on the conception of their possibly cooperative relation with the agent. This conception will mainly be based on their relation during the previous stages in the decision-making process. From an SI point of view, the ratification stage is not the agent's ultimate test anymore, because member states and EU negotiator may share the feeling that they negotiated the MEA 'together'.

4. Research design

4.1. Variables

On the basis of the theoretical framework, the research design can now be built. Two kinds of variables are distinguished: one outcome variable and eight condition variables. In this section, I make use of terminology that is typical of Qualitative Comparative Analysis (QCA), the method that will be used in Chapter 5 to explain the EU negotiator's discretion. An 'outcome variable' corresponds to what is traditionally called a dependent variable, and the 'condition variables' to the independent variables. In other words, the outcome variable is the variable I aim to explain, whereas the condition variables are the explanatory variables. A theory-driven variable selection is a prerequisite for a high-quality QCA procedure (Berg-Schlosser, De Meur, Rihoux, Ragin, 2009). Indeed, the main rationale behind the selection of the variables is that they originate from the theoretical model. Hence, the variables are all derived from the principal–agent model that is extended in three ways. An overview of the variables, as well as their link with the theoretical framework, is presented in Table 3.1.

Table 3.1   Selected variables

Variable code	Name of variable	Kind of variable	Link with theoretical framework
<disc>	discretion	outcome	
<prefprin>	degree of preference homogeneity among the principals	condition	basic model
<prefpa>	degree of preference homogeneity between the principals and agent	condition	basic model
<polit>	level of politicization	condition	basic model
<infag>	information asymmetry in favour of the agent	condition	basic model
<infprin>	information asymmetry in favour of the principals	condition	extension 1
<extcomp>	degree of compellingness of the external environment	condition	extension 1
<agprin>	agent as subset of the principals	condition	extension 2
<instdens>	institutional density	condition	extension 3

Following the definition of 'autonomy' by Hawkins, Lake, Nielson and Tierney, I define the outcome variable *discretion* (<disc>) as the 'range of potential independent action available to an agent after the principals have established mechanisms of control' (Hawkins, Lake, Nielson, Tierney, 2003, p. 4). As discretion cannot only be explained by the behaviour and characteristics of the principals, which in principle steer the agent (Salacuse, 1999), the agent's characteristics have to be taken into account as well. Therefore, I argue that the range of the agent's potential independent action is the result of two distinct – although not completely independent – dynamics, one from the principal's point of view ('granted autonomy'), the other from the agent's ('conquered autonomy') (Delreux, 2009a):

- on the one hand, *principals grant autonomy* to the agent, but they simultaneously establish control mechanisms to monitor the agent and to decrease the likelihood of slack. The net result of delegation minus control is the granted autonomy;
- on the other hand, *the agent may conquer autonomy* because he aims to realize his preferences or he aims to reach an international agreement. It is important to note that conquering autonomy does not automatically refer to agency slack, or unwished agent behaviour.

Measuring discretion may face the researcher with the 'methodological pitfall of observational equivalence' (Damro, 2007). It is indeed possible that indications of autonomous agent behaviour are observed by the researcher, but that this behaviour is in fact not autonomous, but the agent rationally anticipating the control mechanisms of the principals (Pollack, 2003a). To overcome this methodological difficulty, I follow Pollack's and Damro's suggestion to measure the agent's behaviour carefully, e.g. by process tracing and by conducting systematic expert interviews (see Section 1.3 of Chapter 1). I measured discretion by the combination of four indicators. First, delegation of negotiation authority from member states to an agent is a necessary condition for discretion. Second, the likelihood of discretion increases when principals do not take the floor at the international level. Third, the deployment of control mechanisms by the principals decreases the agent's discretion. Finally, an agent conquering autonomy indicates a higher degree of discretion.

The eight condition variables that can explain the EU negotiator's discretion are presented below, as well as the way I measured them. Conditions 1 to 4 are derived from the basic principal–agent model, whereas conditions 5 and 6 are based on the first extension (on the asymmetrical information in favour of the principals and the external compellingness), condition 7 on the second extension (on the mixed nature of MEAs) and condition 8 on the third (on SI).

First, there is a broad agreement in principal–agent literature that the *preference homogeneity among the principals* (<prefprin>), and in particular the extent to which the principals' preferences overlap, can affect the agent's discretion. An EU decision-making process characterized by homogeneous preferences among the principals is theoretically considered to decrease the agent's discretion (Nicolaïdis, 1999; Pollack, 2003a; Hawkins, Jacoby, 2006; Elgström, Frennhoff Larsén, 2010). I measured the degree of preference homogeneity among the member states by combining the perceptions of the decision-makers about the preference distribution, going from minimalist (i.e. not wanting a strong MEA) to maximalist (i.e. striving for a strong and demanding MEA) preferences.

Second, not only the degree of preference homogeneity among the principals can affect discretion, the degree of *preference homogeneity between principals and agent* (<prefpa>) is a potential explanatory factor as well. Determining and defining this variable is difficult in a situation characterized by heterogeneous preferences among the principals. A benchmark, which stands for the 'preference of the principals' and to which the agent's preferences can be compared, has to be defined. I opted to use the pivotal player's preference within the collective principal as benchmark. This is the point where the outcome of a QMV vote can be expected. This means that the preferences of the agent are considered homogeneous with these of the principals if they fall within the majority group of the member states' preferences. If, on the other hand, the EU negotiator's preferences fall outside the scope of the majority of the member states, the preference homogeneity between agents and principals is considered heterogeneous. As a result, the preferences of outlier principals are not taken into account when the degree of preference homogeneity between agent and principals is determined. The degree of preference homogeneity between principals and agent was measured the same way as the first explanatory variable was made operational: the perceived agent's preferences were compared to those of the principals.

Third, the agent's degree of discretion can depend on the *level of politicization* (<polit>) in the EU. A high level of politicization, which means that the issues at the international negotiation table are politically sensitive in the EU, can decrease the autonomy granted to the agent. It is plausible to expect that member states want to keep control over issues that are politically sensitive to them. I used three proxies to assess this variable: the decision-makers' perception about the ranking of the issue on the political agenda, the level at which the EU decision-making is conducted (if decision-making processes and international environmental negotiations are conducted by ministers, they seem to be more politicized than decision-making and negotiation sessions attended by experts or civil servants), and the degree of mediatization of the decision-making process.

Fourth, most principal–agent analyses emphasize an *information asymmetry in favour of the agent* (<infag>). The EU negotiator is well acquainted with the substance and the development of the international negotiations. As the agent can withhold (parts of) this information from the member states, they are uncertain about their negotiator. I measured the degree of information asymmetry in favour of the agent in a twofold way. On the one hand, I assessed the decision-makers' perception of the presence or absence of private information for the EU negotiator. On the other hand, international negotiation sessions that could only be attended by the agent and not by the principals indicate the possibility of asymmetrically divided information in favour of the agent.

Fifth, I not only consider an information benefit for the agent, but also *asymmetrically divided information in favour of the principals* (<infprin>). Member states can possess private information about their fallback positions and the range of agreements they will ultimately be able to accept, which can affect the autonomy they grant to the EU negotiator and the degree to which the agent conquers more autonomy. The data for evaluating this variable was obtained by asking decision-makers whether fallback positions were tabled at the end of the process and whether they assess the agent was informed about the range of agreements that could realistically be ratified.

Sixth, a high degree of *compellingness of the external environment* (<extcomp>) increases the political cost of no agreement for the member states. This makes sanctioning the agent less likely, which in turn increases the likelihood of granting more autonomy. Moreover, it can lead to an agent conquering autonomy, as the international context can *de facto* forces him to go beyond the instructions of the principals, or at least to interpret them broadly. A relatively high number of negotiating partners with a sufficiently large bargaining power (e.g. regional negotiations are considered to be less compelling than global negotiations are), a demanding position of the EU at the international level (i.e. the requirement that something has to be at stake for the EU) and the decision-makers' perception about the pressure experienced by the member states were used as indicators for a high degree of compellingness.

Seventh, the agent's discretion can be influenced by the question whether the *agent is a subset of the principals* (<agprin>). In international environmental negotiations, it is possible that the EU negotiator is not only an agent, but still a principal. While the Commission is not a subset of the principals, the Presidency and a lead country are. As explained in Section 3.2.1, it is plausible to expect that member states that operate as EU negotiator, namely the Presidency or a lead country, will enjoy less discretion than when the Commission acts as agent.

Eighth, *institutional density* (<instdens>), leading to a more cooperative way of decision-making, can also affect the principal–agent relation. To make this variable operational, I measured the decision-makers' assessment of the degree to which the institutional norms of the Council played a role. The occurrence of these norms increases the likelihood of a collective problem-solving atmosphere and an 'administrative partnership' between principals and agent (Joerges, Neyer, 1997). Such a cooperative setting can influence the agent's discretion, although it is not *a priori* clear in which direction. On the one hand, a dense institutional environment can lead to a 'partnership' between principals and agents, which may reduce the gap between the member states and the EU negotiator (and the latter's discretion). On the other hand, it is also plausible to expect an institutionally dense decision-making context to lead to a high degree of granted autonomy. Following this reasoning, the member states show a kind of empathy for the EU negotiator and for the situation the agent is confronted with at the international level. In other words, the principals are likely to grant a high degree of autonomy because the institutional norms will create a cooperative atmosphere, which makes the misuse of autonomy less likely.

4.2. Cases

As the case selection *as such* can have an impact on the outcome of the analysis
– particularly when QCA is used –, this has to be done rigorously (Rihoux, Ragin,
2004). Constructing an area of homogeneity is the main case selection strategy
(Ragin, Berg-Schlosser, De Meur, 1996; Aarebrot, Bakka, 1997; De Meur, Rihoux,
2002). This implies that those cases are selected that are sufficiently comparable
and share enough characteristics (Lijphart, 1975; Sartori, 1994). Such cases form
an area of homogeneity. Here, a case is defined as a 'recent EU decision-making
process with regard to an international negotiation leading to an MEA, which is
signed by the EC and the member states'. To build this area of homogeneity, I started
from a list of all the EU decision-making processes with regard to international
negotiations leading to an MEA, in their its broadest sense. The decision-making
processes in this universe were then filtered by five case selection criteria in order
to arrive at a set of cases corresponding to my definition.

First, the EU decision-making process had to be *recent*. In practice, this means
that the EU decision-making processes must have conducted between 1 November
1993 and 1 May 2004. This guarantees that the legal procedures remain constant
(i.e. the Maastricht, Amsterdam and Nice Treaties, which included the same
provisions on the EU as international environmental negotiator, were in force)
and that there is not much variation in the number of principals (12 or 15). Indeed,
the lower temporal limit is the entry into force of the Maastricht Treaty,[7] while the
upper temporal limit is the Central and Eastern European enlargement.

Second, only EU decision-making processes with regard to negotiations
that are sufficiently *multilateral* are included in the area of homogeneity. This
guarantees a possible degree of external compellingness, which largely depends
on the participation of a sufficient number of 'dominant', 'strong' non-EU states
exerting a high degree of relative bargaining power vis-à-vis the EU.

Third, only decision-making processes with regard to primarily *environmental*
negotiations are selected. To be selected as a decision-making process with regard
to an international negotiation leading to an environmental agreement, these
agreements have to be concluded by the EC on the basis of article 300 TEC *juncto*
article 174 or 175 TEC (see Chapter 2). Moreover, the EU decision-making process
has to be conducted within the institutional context of the Environment Council
and DG Environment was the leading DG in the Commission.

Fourth, the legal status of the result of the international negotiations with
regard to which the EU decision-making process is conducted has to be a *treaty*.
The treaty criterion is applied for two reasons. On the one hand, only treaties
need to be ratified. This is not the case for e.g. COP or MOP decisions. As the

---

7   I allow for one exception to this lower limit criterion: the UN Convention to Combat
Desertification. The first and the second negotiation sessions leading to this Convention
were held in May and September 1993. However, the greater part of the negotiations were
conducted after November 1993.

(anticipation on the) ratification stage is an important element of my theoretical model, ratification has to be part of the decision-making processes studied. On the other hand, only internationally legally binding treaties guarantee that article 300 TEC is followed for the negotiation of the issues falling under EC competences. As my theoretical model is based on the provisions of article 300 TEC, this has to be constant as well.

Fifth, the *EC and the member states should have signed* the agreement with regard to which the EU decision-making process is conducted. Only if the EC signed is the internal decision-making process based on article 300 TEC. For this reason, I do not select decision-making processes with regard to international negotiations leading to an MEA, which is signed by the EU member states but not by the EC, although one might expect – at least informal – coordination among the member states with regard to such negotiations.

The case selection process, based on theoretical and empirical considerations, resulted in eight cases: the EU decision-making process with regard to the UN Desertification Convention (UNCCD), the African-Eurasian Waterbirds Agreement (AEWA), the Kyoto Protocol, the Aarhus Convention, the Rotterdam PIC Convention, the Cartagena Protocol, the Stockholm POPs Convention and the SEA Protocol. They are presented in Table 3.2, together with their code that will be used in this book, the date the agreement was signed, and the date the EC ratified the agreement.

Table 3.2   Selected cases

Case code	Case: EU decision-making process with regard to the international negotiations leading to...	Date of signing	Date of EC ratification
CCD	UNCCD	17/06/1994	09/03/1998
AEWA	AEWA	15/06/1995	18/07/2005
KYOTO	Kyoto Protocol	11/12/1997	25/04/2002
ARHUS	Aarhus Convention	25/06/1998	17/02/2005
PIC	Rotterdam Convention on PIC	10/09/1998	19/12/2002
CART	Cartagena Protocol on Biosafety	29/01/2001	25/06/2002
STOCPOP	Stockholm Convention on POPs	22/05/2001	14/10/2004
SEA	SEA Protocol	21/05/2003	20/08/2008

Each of these MEAs was negotiated in a series of negotiation sessions. Appendix B gives an overview of the various negotiation sessions leading to the selected MEAs. Some of them were negotiated in two sessions; for others, ten or more sessions were needed. In Figure 3.9, each negotiation session for each case is represented on a time axis. Moreover, the moment of signing and which member state held the EU Presidency at that time are indicated.

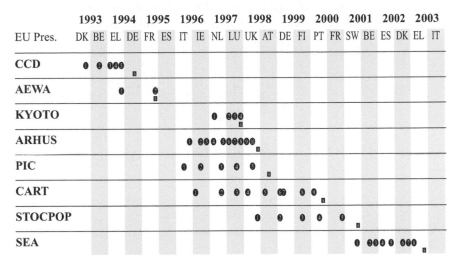

	1993	1994	1995	1996	1997	1998	1999	2000	2001	2002	2003
EU Pres.	DK	BE EL	DE FR	ES IT	IE NL	LU UK	AT DE	FI PT	FR SW	BE ES	DK EL IT

Figure 3.9 **Chronological overview of the negotiation sessions leading to the selected MEAs**

Note: ●: negotiation session; ▧: moment of signing

5. Conclusion

The threefold extended principal–agent model outlined in this chapter frames the EU decision-making process with regard to international environmental negotiations as a relation between member states-as-principals and an EU negotiator-as-agent (basic model), in which the international negotiation context needs to be taken seriously (extension 1), the mixed nature of MEAs matters (extension 2) and sociological institutionalist dynamics may occur (extension 3). On the basis of this model, I developed a research design and I selected the variables that will be used to explain the EU negotiator's discretion. As the empirical data presented in the next chapters will largely be structured on the basis of these variables, they will also come back in Chapters 4, 5 and 6.

I defined six control mechanisms that can be used by the member states to limit the autonomy they grant to their EU negotiator. As the extent to which these mechanisms are deployed is inversely related to the granted autonomy, they will be used to measure discretion (Chapter 4). Moreover, the practical functioning of each of the six control mechanisms will extensively be discussed and their control function will be evaluated. In Chapter 6, the question will be addressed whether the extensions of the principal–agent model correspond to the principal–agent relation as it occurs in reality. Particular attention will be paid to the SI perspective on the principal–agent relation in the EU. In Chapter 7, finally, an evaluation of the applicability of the principal–agent model and its extensions for understanding the practice of the studied EU decision-making processes will be presented.

Chapter 4
The EU as Negotiator in Eight International Environmental Negotiations

1. Introduction: a systematic description of the cases

The aim of this chapter is to describe the eight cases on the basis of the nine variables (one outcome variable and eight condition variables). The eight EU decision-making processes will be presented with a twofold objective. On the one hand, the case studies provide insights in the EU decision-making processes with regard to international negotiations leading to an MEA. On the other hand, the empirical data presented in this chapter serves as the basis for analyses in the next chapters, in which references will be made to the cases. This chapter aims to 'develop an "intimacy" with each of the cases, to gain a deep, "thick" understanding of each case' (Rihoux, Ragin, 2004, p. 10). This is a requisite starting point for a QCA procedure, which will be performed in the next chapter. Quotations of member state or Commission officials, which originate from my interviews, are used to illustrate the case descriptions.

As the aim of the analysis is to systematically compare the cases, a systematic description of each case is needed. Therefore, each decision-making process will be described following the same ten sections:

- in each Section 1, the *international negotiation context*, with regard to which the EU decision-making took place, is portrayed. Presenting the subject matter of the negotiated MEA ('what is at stake?') is consistently followed by an overview of the development of the negotiations and by an outline of the positions of the main players at the international level. Special attention is paid to the EU position;
- in each Section 2, the *EU negotiation arrangement* is described. Which actor(s) negotiated on behalf of the EU during the international negotiations? Or in theoretical terms: which actor(s) was/were the agent(s)? This information is not only needed to understand which actor's discretion will be assessed in a particular case, it also answers the question whether the *agent is a subset of the principals* (<agprin>);
- in each Section 3, the agent's *discretion* (<disc>) is dealt with in two main subsections. The first subsection tackles the question of the autonomy granted by the principals to this agent. To describe the granted autonomy, the extent to which the six control mechanisms were actually deployed (i.e. the potential contestation of negotiation authority, the mandate, the

instruction for the agent, the EU coordination meetings, the participation of the principals at the international level, the risk on an involuntary defection and the ratification process) is discussed. The second subsection touches upon the autonomy conquered by the agent;

- in each Section 4, the *preference homogeneity among the principals* (<prefprin>) is described by defining the preference distribution among the member states, going from the most minimalist to the most maximalist member states;
- in each Section 5, the *preference homogeneity between principals and agent* (<prefpa>) is assessed by comparing the common member states preference to the preference of the agent;
- in each Section 6, the *level of politicization* (<polit>) related to the internationally negotiated issues is presented;
- in each Section 7, an answer is given to the question whether the *agent had an information benefit* (<infag>) vis-à-vis the principals;
- in each Section 8, the possible *information asymmetry in favour of the principals* (<infprin>) is dealt with;
- in each Section 9, the *degree of external compellingness* (<extcomp>) is portrayed; and
- in each Section 10, by systematically describing to what extent diffuse reciprocity, mutual responsiveness, a striving for consensus and compromise, and trust played a role in the EU decision-making process, the *degree of institutional density* (<instdens>) in the EU is considered.

At the end of this chapter, the findings are presented in a summarizing Table 4.2, which makes it possible to compare the different variables of the eight cases in a systematic way. On the basis of this comparison, each variable will be recoded into a binary value to execute the QCA procedure in Chapter 5.

2. UN Convention to Combat Desertification [CCD]

2.1. The international negotiating process

The UN Convention to Combat Desertification is the first international legally binding agreement tackling the desertification problem. It aims to combat desertification and soil degradation in a number of specified regions and to mitigate the harmful effects of drought. The Convention consist of a main text and five regional annexes.[1]

UNCCD was negotiated in five meetings of the Intergovernmental Negotiating Committee on Desertification (INCD) between May 1993 and June 1994, before

---

1   These regional annexes are meant to refine the implementation provisions for respectively Africa, Asia, Latin America and the Caribbean, the Northern Mediterranean (the so-called 'European' Annex 4), and Central and Eastern Europe.

being signed in October 1994. The establishment of the INCD – and thus also the Convention – is generally seen as the result of a political deal between industrialized and developing countries at the 1992 United Nations Conference on Environment and Development (UNCED) in Rio. The desertification issue was only brought to the international negotiation level because this was necessary to keep the developing – and mainly the African – countries on board in Rio and to gain their support for the two other Rio Conventions, namely the UN Framework Convention on Climate Change (UNFCCC) and the Convention on Biological Diversity (CBD) (Toulmin, 1995; Jamal, 1997; Jokela, 2002).

During the INCDs, negotiations were characterized by a split between the G77 on the one hand and the WEOG (Western European and Others Group) countries on the other hand.[2] The G77 was very keen on the Convention that was seen as 'their' part of the Rio process, while WEOG had a more conservative position on desertification. However, as was the case within the G77 (Najam, 2004), not all WEOG countries were in full agreement. The EU was prepared to move into the direction of G77 because it wanted to maintain good relations with the G77 that were needed in other – for the EU more important – international negotiations, whereas the Juscanz countries often immediately blocked G77 proposals.[3] The main question on desertification was whether the problems could be considered as 'global', meaning that the Convention should be financed by a new 'desertification window' in the Global Environment Facility (GEF), which assists developing countries by funding projects and programs that protect the global environment (Toulmin, 1995). As a consequence, the financing question – and more in particular whether the financing of UNCCD would work via the GEF and whether the financial means would be 'new and additional' or a reallocation of existing financial means – was the main point under discussion between G77 and WEOG. The G77 wanted new financial means and a new fund for desertification, while the industrialized countries formed a 'united veto coalition against new financing' (Jokela, 2002, p. 309) and were opposing the use of the GEF for this Convention. WEOG countries considered the traditional development cooperation channels as the most appropriate way to deal with desertification.

2.2. EU negotiation arrangement

The EU negotiation arrangement during the UNCCD negotiations was characterized by a large role for the member states and a rather limited and formal agent role for

---

2 The G77 is a coalition of developing countries, often used as a negotiation bloc in UN negotiations. Nowadays, approximately 130 countries are a member of the G77. WEOG is one of the five regional groups in the UN. It is formally only used for voting purposes. Besides European countries, also Canada, New Zealand and Australia are members of WEOG. The US and Israel are observers.

3 'Juscanz' is usually used to describe a group of countries, consisting of Japan, the US, Canada, Australia and New Zealand.

the Presidency, who acted as the EU negotiator in plenary. UNCCD was negotiated by three different EU Presidencies: Denmark, Belgium and Greece. When the Presidency had expressed the EU position, the floor was open for the member states as well. The member states then mostly followed the general EU line put forward by the Presidency and they gave additional arguments or reworded this position. However, member states could also use this forum to present their own ideas and their national preference, if these diverged from the EU majority position. This only happened occasionally. The fact that member states with positions similar to these expressed by the Presidency took the floor afterwards and repeated these positions, was often an explicit strategy by the EU, since it was considered as more effective that some arguments were presented more than once.

The EU negotiation arrangement in the various small working groups was informal. Unlike in plenary, there was no EU negotiator. These groups were open for any representative and the member states could participate as UN member state. However, in practice, they mostly stuck within the boundaries of the common EU position. In the last stages of the negotiations (mainly INCD 5), the chair of the international meeting called several Friends of the Chair meetings in which eight representatives of G77 and eight representatives of the WEOG could participate (Najam, 2004). From EU side, the (Greek) Presidency and one or two closely involved member states (Germany and sometimes the UK) participated. The Presidency played a very limited role in these negotiations, as it was only present in its formal capacity. As at that time the negotiations were concentrated on the North-South division (WEOG versus G77) and as WEOG tried to act as unified as possible, the Australian WEOG Presidency mostly spoke on behalf of the industrialized countries, including the EU. Consequently, the principal–agent relation between EU member states and their Presidency was replaced by a negotiation arrangement characterized by one non-European agent speaking on behalf of the whole industrialized world.

Although the EC signed UNCCD during the ceremony in Paris in 1994, the Commission did not represent the EC or the member states during the negotiations at the international level. The first reason is that there were not many EC competences in the domain of desertification on the basis of which the Commission could have claimed a role. Second, as the negotiation process was mainly following UN logics and procedures, the Commission was only present in plenary as observer. However, member state representatives acknowledge that the Commission participated in EU coordination meetings in a rather active way.

2.3. Discretion of the EU negotiator <disc>

The agent role of the Presidency was never contested by the member states. 'There was no dispute about the Presidency leading. That was accepted from the beginning' (member state representative). Another member state official even stated that 'there has never been a discussion about the Presidency's role in the negotiation process'. It was, in other words, taken for granted that the member states would participate in this UN process via the EU level, even if there were very few EC competences at stake. The common objectives of the member states – mainly not to provide any additional money in a financial facility of the Convention – was the main incentive to cooperate in the EU, yet also in the broader framework of WEOG. The Presidency operated under clear instructions from its principals, which came down to some very strict points: the EU could never accept the definition of desertification as a 'global problem' or the use of new international funds for UNCCD's objectives.

During the INCDs, EU coordination meetings were organized on a daily basis. However, after each coordination meeting at EU level, another coordination meeting took place with the group of industrialized countries in a WEOG setting. In these WEOG meetings, the EU followed an informal guideline that the Presidency presented the EU position and that the member states could take the floor afterwards, preferably – and also mostly in practice – not diverging from the general EU line. This was a similar arrangement as the one for the INCD plenary. All member state representatives considered the WEOG coordinations, which were also held daily, as even more significant than the EU coordination meetings. As the negotiations were characterized by a North-South divide, the WEOG was a more important coordination body than the EU. The WEOG coordination functioned as the forum to agree on a common 'Northern' voice vis-à-vis the G77.

As the UNCCD negotiations very much followed a UN logic, EU member states could participate in all meetings. Consequently, they were always there when the Presidency took the floor in plenary. However, as already mentioned, by the end of the negotiations, a more limited Friends of the Chair negotiation setting was used. Not every EU member state could attend these meetings. However, it was not only the Presidency that negotiated on behalf of the EU in this setting. First and foremost, the agent role was mostly taken up by Australia, chairing the WEOG. Second, closely involved member states (Germany and sometimes UK) were also part of the eight-headed WEOG delegation. During the five INCDs, there were no member states threatening that they would not sign or not ratify the Convention. Neither the ratification in the Council in 1998 (98/216/EC) nor the different ratification processes at the national levels posed any problems to the EU.

Member state representatives admitted that the Presidency neither conquered more autonomy nor went beyond the instructions given by the member states. In the final stages of the negotiations that were only attended by a limited number of WEOG and G77 representatives (i.e. Friend of the Chair negotiations) member states were constantly debriefed, mainly through WEOG channels (and not anymore via the EU). The negotiations were frequently interrupted to allow the negotiators to go back to their group (the G77 representatives had to go back to their group as well). No decision was taken in a Friends of the Chair setting without the backing of the principals. Consequently, because of intense consultation processes between principals and agents, there was never a problem with a WEOG or EU negotiator going beyond his instruction or conquering additional autonomy. Moreover, the final agreement was reached in plenary, where all member states were present, although it was to a large extent prepared in the Friends of the Chair setting.

2.4. Degree of preference homogeneity among principals <prefprin>

Generally speaking, there were no fundamental disagreements on UNCCD within the EU. The member states shared a general approach on the Convention: they admitted that desertification is a problem in certain regions, but they were convinced that an MEA in which the problem would be addressed as 'global' was not the most appropriate way to deal with the problem. The EU considered development cooperation as a more appropriate answer to the desertification problem than an environmental agreement. There was a low degree of variation among the principals' preferences. While member states like the UK and Germany held a very strict minimalist line,[4] the Netherlands (because of the profiling of the Dutch Development Minister on the issue) and France (because of its traditional link with the West African countries) had somewhat more moderated preferences. However, despite these little differences, the degree of preference homogeneity can be considered as high.

There was one issue on which the member states' preferences varied: UNCCD's Annex 4. This annex refers to desertification in the Mediterranean member states Spain, Portugal, Italy and Greece. In particular Spain, being the pioneer of this Annex, wanted to have included a 'European' Annex to UNCCD, in which the desertification issues in the Southern member states would be addressed. As a result, the position of these principals tended somewhat more towards the maximalist G77 position. A member state official portrayed the position of these member states as follows: 'They were just more sympathetic to the G77. They would look for opportunities to be supportive to the G77 without diverging from the main EU negotiating line.' The main reason for Spain – probably silently supported by Italy, Portugal and Greece – to ask for that Annex was not to get money out of a financing system connected to UNCCD, but to try to use recognition as a country affected by desertification as a lever for a spill-over effect to the EU level, from which they hoped they could get financial means from the EU budget. At INCD 3, the other EU member states agreed with a fourth Annex, as long as it would not have financial consequences. From the moment that Spain had won its point,[5] the preference distribution in the EU can be considered as completely homogeneous. For this reason, this issue may not be seen as so dominant in the EU decision-making process that it really threatened the overall consensus.

---

4   These member states with minimalist preferences opposed an international agreement on desertification as such. However, they were aware that they could not stop the establishment of UNCCD because it was linked to other Rio processes. From that point of view, they strived for a convention with rather weak provisions.

5   However, Annex 4 is described as 'weak' and 'meaningless' by member state representatives.

2.5. Degree of preference homogeneity between principals and agent <prefpa>

In the period of the UNCCD negotiations, three member states held the Presidency in the EU: Denmark, Belgium and Greece. There were no preference conflicts between the negotiating Presidency and the member states. None of the Presidencies had a substantial national interest in the Convention. The Danish and the Belgian preferences were situated in the middle of the small range of principal preferences between the UK and the Netherlands, although Belgium was maybe somewhat more minimalist than Denmark and a bit closer to the strict UK preference. Most member state officials could not define the Greek preference and stated that Greece 'had no position'. Hence, the agents' preferences were situated within the range of the principals' preferences and the degree of homogeneity between both was high. Member state representatives portrayed the three Presidencies as 'neutral'.

2.6. Level of politicization <polit>

The UNCCD negotiations were not politically sensitive in the European member states. Except for the Spanish plea for a Mediterranean Annex, there were no national interests at stake in the EU. Member states considered this negotiation process as a political commitment towards the G77 to ensure their support for the climate and biodiversity conventions. A member state representative portrayed the situation in his country like this: 'If I had come back to my capital and I would have said "the negotiations have failed", no one would have worried'.

2.7. Information asymmetry in favour of the agent <infag>

The member states could have the same information as their negotiator during the UNCCD negotiations, because they were also sitting at the UN table. Hence, there was no information asymmetry in favour of the agent. Member state officials acknowledge that 'there was a very good feedback from the Presidency'. However, the main source of information for the member states were the WEOG meetings. The fact that WEOG played such a prominent role in these negotiations and in the transmission of information meant that it was impossible for the Presidency to withhold information.

2.8. Information asymmetry in favour of the principals <infprin>

The main players in the EU decision-making process (mainly UK, Germany and France) were clear from the beginning of the negotiations on their position and the range of possible agreements they were able to accept. Member states, which were less involved or which did not have a specific position on the issue, neither enjoyed an information benefit vis-à-vis the Presidency. A member state representative expressed the lack of member states' fallback positions like this: 'Because the political importance was very limited for the member states and because there were no national interests at stake, the negotiations in the EU were very open'.

2.9. Degree of compellingness of the external environment <extcomp>

As the member states were a demanding party in these negotiations, the UNCCD negotiations did not generate a large degree of compellingness towards them. However, some national officials noticed a kind of pressure, as a failure of the negotiations and being held responsible for such a failure would have had consequences for other environmental negotiations (*in casu* on climate change and biodiversity), which were considered as much more important. If there was some pressure on the member states, it came from the negative effects a possible failure would have had.

2.10. Institutional density <instdens>

The EU decision-making process during the INCDs can be described as 'peaceful, friendly and cooperative' and as 'not antagonistic' (member state representatives). There were no tensions or conflicts between principals and the agent. The member states constantly acted in a cohesive way. The fact that member states were continuously trying to accommodate even the most outlier position of Spain on the Annex 4 indicates a certain degree of diffuse reciprocity. Mutual responsiveness logics can hardly be found in this decision-making process, as the member states' positions were clear, homogeneous and not based on a strong national interest. There was no need for a member state to justify its position vis-à-vis the other principals.

During the first INCDs, a compromise was mostly found in the EU, with the exception of the Annex 4 issue. Even on that rather difficult issue, 'if one member state [Spain] did not follow the consensus, we went on. Consensus is not unanimity. Because of their isolation on that issue, they were often ignored in the EU position' (member state representative). However, the striving for consensus was not that prominent: if one member state (Spain on Annex 4, but in the beginning also sometimes the Netherlands on the financing issue) could not agree on an EU position, the position of the other 11 member states was presented by the Presidency. In the international negotiations, the 11 member states supported the EU line, while the member state with the minority position could freely express its position at the international level as well.

Finally, the Presidency was trusted by the member states as their negotiator at the international level. A member state official portrayed the EU coordination meetings as 'an atmosphere of trust'. The people around the EU decision-making table mostly knew each other since a long time, as they were involved in the Rio process, of which UNCCD was a result.

3. Agreement on the Conservation of African-Eurasian Migratory Waterbirds [AEWA]

3.1. The international negotiating process

The African-Eurasian Migratory Waterbirds Agreement protects 235 species of waterbirds migrating in the African-Eurasian corridors and for that purpose relying on wetlands in their annual passage. AEWA is concluded under the Convention on the Conservation of Migratory Species (CMS or Bonn Convention, 1979). CMS makes it possible to negotiate regional agreements to protect the species listed in Annex II of CMS, including waterbirds migrating on African-Eurasian passages. AEWA is an intercontinental agreement, negotiated by African, Asian and European parties. The decision to negotiate an agreement was taken at the first COP of the Bonn Convention in 1985, which was followed by a long preparation period that could hardly be considered as a 'negotiation process'. This technical consultation process, in which mainly scientists participated, did not take place in a formalized or institutionalized international context. In that period, the Netherlands, which is nowadays seen as the pioneer of AEWA, played an important role. The process was finalized by a real negotiation session in The Hague in 1995 (Lutz, Boye, Haupt, 2000). One year before, another conference – which was more informal and more consultative than the meeting at The Hague, but which can nevertheless be seen as a negotiation session – had taken place in Nairobi. In the framework of this book, the focus is mainly on the negotiations of 1994 and 1995.

The EU can be regarded as the leading and demanding party in the negotiations. The conservation of (migratory) birds was already regulated in the EU, mainly by the Bird Directive (79/409/EEC) and the Habitat Directive (92/43/EEC). However, the EU's aim was not to copy the Bird Directive to the international level. This was considered as an impossible or an unreachable ambition, in view of the low level of bird conservation in the African and Asian countries. The purpose was rather to ensure that what the EU was doing on bird protection in the Community was not destroyed outside its territory. For the EU, the new idea of the Agreement was not the direct protection of birds, but the close cooperation with African countries for effectively realizing this protection. As the protection and the hunting of birds are closely connected, one of the main issues in the AEWA negotiations – mainly because of the role and position of (some member states of) the EU – was hunting. Hunting NGOs were opposed to the development of AEWA, as they were afraid that the international agreement would impose even more hunting restrictions than those resulting from the Bird Directive, and now with a wider geographical range.

3.2. EU negotiation arrangement

The Commission was the EU negotiator in the AEWA negotiations. However, the agent role of the Commission has to be qualified in two ways. First, the Commission was only the agent in the final negotiation session in The Hague. In the 1994 meeting in Nairobi, member states could play a larger role. An observer described the meeting in Nairobi as follows: 'Everyone participated in the discussions and everybody just talked informally'. Second, although most AEWA issues were covered by Community competences, member states could also take the floor for issues, which did not have a direct impact on the Bird Directive, such as the management plans or the organization of the secretariat and the various scientific committees. In general, it was agreed before in a coordination meeting on which issues the Commission would speak. If the Commission had put forward the EU position at the international level, member states never opposed it: 'Member states never said something in public that contradicted what the Commission said' (member state representative). However, in the coordination meetings afterwards, member states certainly made clear their position if it diverged from the position the Commission had expressed. On the issues on which the Commission was not the agent, the member states were free to take the floor. A member state representative portrayed this situation like this: 'We waited in the discussions whether the Commission representative would raise his voice on a certain issue. If there was no comment from the EU, our delegation decided to make a [national] statement.' Although they were in a position to present their own preferences, member states generally stuck to the general EU line and they did not come up with something that was not discussed before at the EU level.

3.3. Discretion of the EU negotiator <disc>

The AEWA negotiations were not characterized by a large degree of discretion for the Commission. In the ten-year period 1985-1995, there was only one week – the last session in The Hague – in which the Commission was really the EU negotiator. Even when the negotiation arrangement with the Commission as agent was applied, member states could still play their role internationally.

The preparatory stage of the negotiations was not characterized by a EU driven process. By the end of the 1980s, the Commission did not seem enthusiastic about the project of negotiating an intercontinental birds agreement that could possibly reopen the discussions on the Bird Directive internally. Moreover, at that time, the Commission was reserved on the establishment of new international conventions under the Bonn Convention, as this generated bureaucratic and institutional overload, costing a lot of – mainly European – money. Only when the final negotiation meeting in The Hague approached, the Commission claimed a larger role and this was not contested by the member states, since they acknowledged the Community's competences on AEWA. 'As the connection with the Bird Directive was very clear, it was almost inconceivable that an individual member state would negotiate in its own right' (member state representative). Before the meeting at The Hague, the Commission was authorized by the Council to negotiate. This authorization went together with a mandate, strongly stressing the fact that the final agreement must be compatible with the Bird Directive.

During the final negotiation session, EU coordination meetings were organized on the spot and the AEWA issue was on the agenda of a limited number of Council Working Groups in the months before. Except for the hunting issue, member states easily managed to reach a common position, being that AEWA could not deviate from the Bird Directive.[6] Within the limits of the existing EU legislation, certain flexibility was granted to the Commission. All international negotiations were conducted in a setting with open participation. As a consequence, there were no limited Friends of the Chair-like negotiation settings, and member states could attend all negotiations.

In the last days of the negotiations, an incident created by Belgium threatened to jeopardize the negotiations and to place the Commission at risk of an involuntary defection. Belgium requested that its Regions could co-sign the Agreement, as the competences within Belgium were attributed to the federated entities. Member states like Spain, Germany and the UK – being sensitive for federalist issues – declared that they would leave the negotiations and not sign the Agreement if the Belgian request would be granted. To remove this involuntary defection risk, the issue was transferred to the level of the Belgian ambassador in The Hague – the rest of the negotiations being held at expert and civil servant level – who finally agreed not to have a regional co-signature. Whether or not France would accept the Agreement and ratify it afterwards was unclear during the negotiations. France had a very minimalist position on hunting, which dominated the EU decision-making process and even the international negotiation process. Nowadays, not all players in the EU assess the French situation with regard to a possible involuntary defection equally. Some said that France used the argument not to ratify AEWA if the Agreement did not accommodate more its hunting preferences, while others stated that 'there was no direct threat by France' or that it was 'rather to give a signal to their domestic hunting movement and their national public opinion' (member state representatives). In any case, it took the EC nearly eight years to ratify the Agreement (2006/871/EC). This is mainly due to delays in the Commission, since the Commission proposal for ratification dates from 2004 (COM(2004) 531). Hence, it was not a disagreement among the principals, but mainly delays on the side of the agent, that caused a late EC ratification. At the national level, from the 12 member states, only Greece has not (yet) ratified AEWA.

The Commission did not strive to conquer more autonomy than the principals already had granted. All EU decision-makers affirm this statement by a member state representative: 'The Commission always behaved very correctly. They always waited for any comments from member states and were very open to the opinions of the member states. They were always fair. I had very positive experiences with this cooperation. I was never disappointed.' That all member state representatives acknowledge that the Commission constantly stayed in touch with them also indicates that the Commission did not conquer additional autonomy in the AEWA negotiations.

---

6   As member states' positions on hunting deviated (see further), the coordination meetings did not result in a common position on hunting.

3.4. Degree of preference homogeneity among principals <prefprin>

On the substance of AEWA – i.e. on the protection of waterbirds and nature conservation –, member states' preferences were rather homogeneous. However, the major differences in the EU on the hunting issues, combined with the fact that hunting was one of the key issues in the negotiations, made the overall degree of preference homogeneity among the member states very low. The whole discussion often stuck on the French hunting position, which was very minimalist and which made France isolated in the EU.[7] In France, 'where hunting is such an inherent part of their culture' (member state representative), hunting organizations were very much opposed to AEWA because they feared it would become a 'second Bird Directive', imposing even stronger hunting limitations. The pressure of the French hunting organizations certainly influenced the French national position. Moreover, France was not satisfied with the hunting provisions in the Bird Directive, as it wanted to extend the permitted hunting season. By introducing such provisions in AEWA, France hoped to reopen the Bird Directive discussions at the EU level. This was something that the other member states and the Commission tried to prevent at all costs. The Netherlands and Germany were the countries that most actively opposed the French position. Also the UK, which is traditionally keen about nature conservation, was on the maximalist side. These countries were hostile to hunting and already had a restrictive hunting policy.

As a consequence, the principals' preferences were heterogeneous, although this was only due to the hunting issue, since there was less internal EU discussion on the other substantive points of AEWA. However, hunting was so dominant in the negotiations that it overshadowed these other issues.

3.5. Degree of preference homogeneity between principals and agent <prefpa>

As already mentioned, the Commission did not take a leading role in the EU decision-making process with regard to the AEWA negotiations. An observer to the negotiations stated: 'The Commission was not terribly enthusiastic about AEWA'. Since the Commission's priority was that the Agreement did not violate the provisions in the Bird Directive, the agent's preference on AEWA did not diverge from the common position of the member states. Moreover, it was more important for the Commission that the member states first accurately implemented the Bird Directive (which was not yet the case in the beginning of the 1990s).

---

7   As not every member state was active in the EU decision-making process on AEWA, it is difficult to observe real isolation with regard to the French position. Some officials suppose that Spain's preference on hunting could have been rather close to the preference of France, but as Spain was less active and as it almost never took the floor in the EU coordination meetings, it is fair to say that France was *de facto* the only member state hindering an otherwise easily reached common EU position.

3.6. Level of politicization <polit>

Generally speaking, the issues of AEWA were not politically sensitive. There was, however, one exception: the hunting issue in France. But in leading countries like Germany, the UK and certainly the Netherlands, which nonetheless invested a lot of (political) capital in the negotiations, the AEWA negotiations were not salient. There were neither ministerial segments in the negotiations nor much media attention. However, the negotiations were characterized by a rather large involvement of NGOs, on both the hunting and the nature conservation sides.

3.7. Information asymmetry in favour of the agent <infag>

It is generally admitted that, during the negotiations on AEWA, the EU negotiator did not enjoy an information benefit vis-à-vis the member states. The principals attended all international meetings. A member state representative declared: 'I don't have the feeling that the Commission withheld information. The Commission was also open on the information it got in talks with African and Eastern European countries. This was always discussed in the coordination meetings. We got all the information.' Moreover, member state officials acknowledge that the Commission did not use its information strategically. Even the documents for the coordination meetings in Brussels were not prepared by the Commission, although it had the right of initiative and it was the EU negotiator, but by some leading member states.

3.8. Information asymmetry in favour of the principals <infprin>

Except for some indications in the French position, the principals did generally not have information about their preferences or fallback positions that was not known by the agent. 'Inside the EU coordination meetings, the member states were rather frank' (member state representative). In the negotiation stage, the French position was somewhat more difficult to assess, as it was not completely clear how much France could accept on hunting. They often came up with minimalist positions, but it was not always unambiguous whether France could not go further or whether their position was more rhetorical vis-à-vis their domestic level. In other words, the Commission had probably no clear view on the final French win-set.

3.9. Degree of compellingness of the external environment <extcomp>

As the negotiations on AEWA were intercontinental and as the EU and its member states were rather dominant, the external environment was not that compelling. According to a member state representative, 'the EU dominated the AEWA negotiation process. However, I would not go that far to say that the EU dictated the Agreement.' As a consequence, the principals did not really experience pressure from the African and Eastern European negotiation partners. If member states felt pressure, it came from their domestic level (in particular in the case of France) or from the NGOs.

3.10. Institutional density <instdens>

The AEWA negotiations – and in particular the final meeting in The Hague – were 'an example of member states working together in a very good way' (member state representative). Member state and Commission officials admit that the diffuse reciprocity norm played a role in making the EU decision-making process less conflictual than hard bargaining: 'Member states were very well aware of what could go wrong if one would stick to his position too long. There was a kind of courtesy to make the best of it together' (member state official). The EU decision-making process was characterized by an atmosphere of mutual responsiveness as well: 'There was a willingness to help each other to make sure that we had a unified EU position. Member states also learned from each other: the actions plans taken up in the Agreement are a good example where we listened to each other's needs' (member state representative). Also France was constantly justifying its position, although other member states did not give in on that.

Possibly the best indication of a rather dense institutional environment in the EU is the fact that, although the range of member states' preferences was heterogeneous, there was always a consensus and a common position that could be expressed at the international level. All representatives perceived France being isolated in the EU on the hunting issues. A decision-making process characterized by an isolated principal on the one hand and a single voice outcome on the other hand shows that the consensus and compromise striving in the institutional setting of the EU played a role. Even if there were intense discussions in the EU, there was always a desire to solve the disagreements and to reach a common position 'without creating a schism in the EU' (member state representative). There has never been (a threatening of) a vote.

The final institutional norm that points to institutional density, trust, was less present in the AEWA negotiations. As there had been a reshuffle in the nature conservation unit of the Commission, the Commission representative in The Hague was new in the dossier. This led to the principals 'not completely trusting' (member state representative) their agent. Moreover, the style of the Commission team in the EU coordination meetings seemed to be perceived by some member states as 'quite arrogant'. A member state official described the relation with the Commission as follows: 'I think that their way of presenting things suggested that there was maybe a hidden agenda, which did not create trust. However, a hidden agenda has never turned out. And yet, you have that feeling.'

4. Kyoto Protocol [KYOTO]

4.1. The international negotiating process

The 1997 Kyoto Protocol is a Protocol under the United Nations Framework Convention on Climate Change (UNFCCC), one of the three so-called 'Rio Conventions' (1992). UNFCCC aims to stabilize greenhouse gases in the atmosphere to combat the enhanced greenhouse effect, which causes climate change. However, UNFCCC includes neither mandatory limits or emission reductions nor enforcement provisions. The Kyoto Protocol is the first international legally binding instrument to realize the objectives of UNFCCC. Its main elements are quantified reduction targets on the one hand and flexible mechanisms to achieve them on the other hand.

First, the Kyoto Protocol stipulates quantified emission reduction targets for industrialized countries (the so-called 'Annex I countries'[8]) below their 1990 emission levels for the period 2008-2012. Article 4 of the Kyoto Protocol allows the EU to fulfil its 8% emission reduction commitment jointly and thus by differentiated targets for the member states (the so-called 'EU bubble' concept) (Oberthür, Ott, 1999; Damro, Luaces, 2001; Chagas, 2003). Second, the Kyoto Protocol establishes three kinds of flexible mechanisms, which allow the Annex I countries to meet their targets in a market-driven way, i.e. by purchasing their emission reductions elsewhere: emissions trading, joint implementation and the Clean Development Mechanism.

The main negotiations on the Kyoto Protocol were held in a one-and-a-half-week negotiation session in Kyoto in December 1997 (COP 3 of UNFCCC). This final negotiation session was prepared in various Subsidiary Bodies of UNFCCC in the course of 1997, taking place in Bonn. UNFCCC's COP 1, held in Berlin in 1995, had resulted in the so-called Berlin mandate, stipulating that the Annex I countries had to establish quantified emission reductions in a legally binding instrument by the end of 1997. As the deadline of the Berlin mandate came closer in Kyoto, the final days of the negotiations were very intense. The final negotiations on the reduction targets were mainly conducted by the three main industrialized players: the US, Japan and the EU.

Among the Annex I countries, the EU is considered as the leader and frontrunner in the negotiations, taking the most demanding position (Sjöstedt, 1998; Damro, Luaces, 2001; Gupta, Ringius, 2001; Bretherton, Vogler, 2003; Damro, 2006; Schreurs, Tiberghien, 2007), while the US took the most minimalist one.[9] Defending the EU bubble concept (to which Juscanz countries were initially opposed), a binding emission reduction target of 15% in 2010 and the striving for 'common and coordinated policies and measures' to reach the targets took a central place in the EU position (6309/97; 9132/97; 11332/97; COM(1997) 481).

In March 1997, i.e. nine months before the Kyoto negotiation session, European environment ministers agreed a political deal on a Burden Sharing Agreement, in which the (provisional) commitments of the different member states were laid down (6309/97). The Burden Sharing Agreement and the EU bubble are necessarily

---

8   This refers to the Annex I of UNFCCC, which lists 40 industrialized countries, including the European Community and its 15 member states of 1997.

9   If the non-Annex I country positions are taken into account as well, they were even more demanding than the EU. However, non-Annex I countries did not have to make legally binding commitments in Kyoto, according to the Berlin Mandate.

linked to each other, as the international commitment, stipulated in the bubble, can only be fulfilled by an intra-EU division of emission reductions. Although the Council conclusions mention a flat target of -15% for all the parties to the Protocol, the burden sharing only came down to 9.2% emission reductions in the EU,[10] which were divided across the member states, taking into account the different domestic economic and energy situations.[11] However, as the Kyoto Protocol only stipulated 8% reductions for the EU as a whole, the Burden Sharing Agreement had to be renegotiated in June 1998 (09402/98)[12] (Sjöstedt, 1998; Oberthür, Ott, 1999; Lacasta, Dessai, Powroslo, 2002; Chagas, 2003; Pallemaerts, 2004).

The US position, on the contrary, can be characterized as 'free market environmentalism' (Damro, Luaces, 2001), as the US favoured an emission trading system and the Clean Development Mechanism, and opposed fixed targets.[13] The US preference was to stabilize CO_2 emissions by 2010 (i.e. 0% reductions), and only if the developing countries committed to emission reductions as well. Japan's initial position was a reduction of 5%, while Australia even strived for an increase in its emissions (Newell, 1998). The position of the developing (non-Annex I) countries mainly came down to ensuring that their economic and social development was not hindered by environmental efforts to solve a problem that was basically created in the industrialized countries (Yamin, 1998).

4.2. EU negotiation arrangement

While the Dutch Presidency was the EU negotiator in the first half of 1997, the troika setting – meaning here the current, former and future Presidency[14] – was used

---

10   The 9.2% reduction is only the result of a weighted sum of the reductions in the Burden Sharing Agreement. The EU's goal and position still was to reach a reduction of 15% for all Annex I parties in Kyoto (6309/97). Hence, the EU proposed a 15% reduction, but it had only divided 9.2% among its member states.

11   Portugal (+40%), Ireland (+15%), Greece (+30%) and Spain (+17%) got a target that was higher than their 1990 levels, while member states like Denmark, Germany, Austria (all -25%) and Luxembourg (-30%) committed themselves to strong reductions. Belgium, the Netherlands and the UK had to reduce 10% of their 1990 emissions under this agreement, while Italy had to reduce its emissions by 7%. The emissions in France and Finland had to stay at 1990 levels and Sweden was allowed to increase its emissions by 5%.

12   The Burden Sharing Agreement negotiations of June 1998 were characterized as very hard bargaining among the member states. Contrary to the March 1997 Burden Sharing Agreement, the one of 1998 was going to be legally binding for the member states. It definitely became legally binding in 2002, together with the EC ratification decision in 2002 (see further). In principle, it would have been possible to convert the commitments of March 1997 in function of the new situation created by the Kyoto Protocol by means of the rule of three. However, this has not been done, as a result of which member states' 1998 commitments differ from their 1997 political ones.

13   As is known, the US did not ratify the Kyoto Protocol.

14   This was the composition of the climate change troika in 1997. Nowadays, the troika has a different composition in climate change negotiations: the current Presidency,

to represent the EU in the second half of 1997, including in the main negotiation session in Kyoto. In practice, this meant that the Netherlands was the EU negotiator in the first six months of 1997, and that Luxembourg (as the Presidency), the Netherlands and the UK formed the troika in the second half of 1997.

The troika arrangement was used because of the multitude and the complexity of the issues and the negotiation groups, and because the Luxembourg delegation had an insufficient action capacity to negotiate everything. In contrast to the Dutch Presidency, which was considered 'a very strong Presidency' (member state officials) and which invested a lot of political capital, leadership and expertise in the process and in the 1997 Burden Sharing Agreement (Kanie, 2003; Schreurs, Tiberghien, 2007; Vogler, 2009), the Luxembourg Presidency was rather limited in terms of people and vigour. This was an incentive to modify the EU negotiation arrangement and to use the troika as agent.

When the troika functioned as EU negotiator, its three members did not use a predetermined division of labour. However, due to personal experiences of the negotiators and to the interests of the troika member states, a country *de facto* took the lead on a certain issue (e.g. the Netherlands did a lot of work on the policies and measures, UK on flexible mechanisms). During the last three days of the Kyoto meeting, the negotiations were conducted at ministerial level. Consequently, the Luxembourg, Dutch and UK environment ministers took the role of agent in the endgame of the negotiations, which still had to cut the difficult knots. Some Commission and member state officials acknowledge that the UK environment minister John Prescott – also being deputy Prime Minister and thus most senior – strongly took the lead in the troika and that, as a result, the UK played a dominant role in the final days of the negotiations, particularly when it came to bilateral negotiations with the US.

The common EU position was generally expressed by the troika.[15] Member states neither took the floor in plenary nor did they in the various contact groups. This was certainly the case in Kyoto, with an exception every now and then, when the troika asked the member states to support the EU position after it had been presented by the EU negotiator. In the various Subsidiary Body negotiations before December 1997, some active member states[16] spoke at the international level, although this was very limited as well. A member state representative stated on this: 'If member states would have taken the floor in plenary or in the contact

---

the future Presidency, and the Commission.

15   In this sense, my findings diverge from research by Sjöstedt, who writes that 'the EU has clearly not performed fully as a unitary actor in the climate negotiations to the same extent as it has in numerous trade talks with third countries for instance' and that 'the EU demonstrated a split position in the climate talks' (Sjöstedt, 1998, pp. 236-237). However, my findings correspond with these by Oberthür and Ott, who state that '[...] the EU largely succeeded in appearing as a unitary actor during the Kyoto process [...]' (Oberthür, Ott, 1999, p. 17).

16   Examples of more active member states were Germany, France and Denmark.

groups, it would have been considered as a serious fault.' As a consequence, EU member states were never able to promote their own national preference at the international level.

The Commission did not negotiate at the international level. However, it was quite active at the EU level in assisting and supporting the troika. It prepared e.g. papers about the situation in the EU with regard to greenhouse gas emissions, or about the technical feasibility of the EU proposals and their economic justifications. The input by the Commission did not only provide the member states and the EU negotiators with the necessary background information, it was also instrumental in the negotiations with the external partners as these papers for instance showed that the EU would be able to reach the targets it was proposing (Pinholt, 2004). Although the EC is a party to UNFCCC and the Kyoto Protocol, article 300 TEC procedures were not followed, not even partially. This can be explained by three reasons. First, in 1997, EC competences in the area of climate change were still very limited. Second, member states were concerned that the Commission would try to increase EC competences on climate change issues and that it would use its active participation in the Kyoto negotiations to claim EC competences in the implementation of the Protocol afterwards. A member state official described the relation with the Commission as follows: 'It was a constant tension between the concern of giving the Commission too much competence on the one hand and to bring the Commission aboard to help us on the other hand.' Third, a single (and strong) EU voice was in the interest of the Commission as well, as this strategy was considered necessary to get the EU bubble concept included in the Protocol. The Commission assumed a large role for itself in the implementation and enforcement of the Burden Sharing Agreement when this would come into force. As a consequence, the Commission seemed to have reasoned that they could better leave the formal negotiating role to the Presidency (or the troika) and play an influential role behind the scenes at the EU level.

4.3. Discretion of the EU negotiator <disc>

The Presidency's and troika's negotiation authority were not contested by the member states. There are three reasons for this. First, a single EU voice was considered crucial to obtain the EU bubble and it was taken for granted that this single voice would be expressed by the Presidency, if necessary assisted by its predecessor and successor. Moreover, using this strategy, by which member states could hide themselves behind the EU bubble, meant that the member states were able to postpone a decision on their own national targets. In other words, striving for one emission reduction target for the EU as a whole not only triggered the single EU voice at the international level, but it also allowed the member states not to take a decision at the domestic level about their national targets. Second, all member state representatives confirmed the bargaining power argument by stating that 'in these negotiations, it was certainly a strategic benefit to negotiate as a block because that is the only way to carry much weight', or that 'it was all about having impact at the international level'. Third, already during UNFCCC's first two COPs, the EU had used its single voice strategy. As a result, this was already a kind of established practice. However, there was no formal authorization decision.

The instructions for the EU negotiator were not put in a formal negotiating directive or a mandate. However, the conclusions of the 1997 Environment Councils – and mainly the one of March 1997 (6309/97) – can be considered as the basic negotiation instruction. These Council conclusions formed the basis for EU position papers, which were established for each article of the draft Protocol, and which ultimately led to speaking notes for the troika members.

However, the authority granted in the instructions broadened when the negotiations came to their endgame. Leaving Kyoto without an agreement was something the EU and its member states definitely wanted to avoid. The EU, which had started the negotiations as the frontrunner among the Annex I countries, strived for a compromise in Kyoto. This is the main reason why the member states ultimately gave in on some of their strict positions laid down in the Council conclusions, e.g. on the emission trading system. The member states were initially not in favour of flexible mechanisms, yet they allowed their negotiator to agree on it because this seemed to be the only way to get fixed and quantified reduction targets in the Protocol. That the EU could only move on these points at the end of the negotiations, can be explained by the fact that these positions were laid down in Council conclusions. As these conclusions were adopted at ministerial level in Brussels, they could only be modified by ministers, who were only in Kyoto during the final days.

While the instruction can generally be considered as quite broad, it was very small on some particular points: certainly on the EU bubble, but also on the targets. Moreover, not everything could be covered by the Council conclusions, as some of the main points of the Kyoto Protocol only came on the table during the Kyoto meeting itself, the Clean Development Mechanism being the most prominent example. Giving precise instructions to the agent worked well in the Subsidiary Body negotiations and in the first days in Kyoto. However, when the negotiations intensified in the endgame of the negotiations, the member states were no longer able to provide their agents with detailed instructions. A Commission representative described this situation as follows: 'When we came in the last rounds of the negotiations, it went too fast to have EU positions on paper'.

Many EU coordination meetings were organized, both before in Brussels and on the spot in Bonn and Kyoto. Particularly in Kyoto, the EU coordinated on a quasi-permanent basis. Moreover, often multiple expert coordination meetings were organized simultaneously. In the final days in Kyoto, the EU held an almost non-stop coordination meeting. This was necessary to react to new proposals coming from the international level. During these days, the Protocol was negotiated in a Friends of the Chair setting on the one hand and in a limited negotiation setting, which mainly dealt with the targets, with the EU, the US and Japan on the other hand. In both settings, the EU was only represented by the troika, without any other member state being present. This generated some tensions between the agents and the principals because some member states feared that they were not sufficiently involved. This feeling was even strengthened because of the presence of ministers, who wanted to be involved.

None of the Commission or member state officials could recall a situation where a member state had threatened to leave the negotiations, not to sign the agreement or to block its national ratification or the ratification in the Council. In other words, there was never a risk that the agent would be faced with an involuntary defection. However, the EC ratification cannot be considered as a formality. In the ratification stage, member states began to recognize that reaching their targets would not be easy. Moreover, member states did not want to ratify before some provisions of the Kyoto Protocol (e.g. on sinks and on the flexible mechanisms) were concretized. After these operational details were agreed upon in 2001 (COP 7 in Marrakech, the so-called Marrakech Agreements) (Dessai, Schipper, 2003), the Council ratified the Protocol (2002/358/EC). This ratification decision differs from other MEA ratifications, as it not only includes the EC's consent to be bound, but also additional provisions for the internal realization of the joint EU commitments (Pallemaerts, 2004). Indeed, the Burden Sharing Agreement of June 1998 legally entered into force by this ratification decision, as the division of emission reduction targets among the member states are incorporated in the second annex of decision 2002/358/EC. The Kyoto Protocol is ratified on the basis of article 175§1 TEC, stipulating QMV. However, some member states aimed, because of the Protocol's possible consequences on energy policy, to base this decision on article 175§2 TEC, implying unanimity.[17] This disagreement on the legal basis was solved by keeping article 175§1 TEC as the substantive legal basis, but to include the unanimous character of the decision in the minutes of the Council meeting (Pallemaerts, 2004). Moreover, the eleventh consideration of the ratification decision mentions that the legal basis of this decision does not establish a precedent for concluding future international agreements on emission reductions with QMV. At the national levels, all member states ratified the Protocol in 2002.

Member state officials admit that, during the negotiations in the Subsidiary Bodies and in the first week of the Kyoto meeting, the Presidency and the troika did not go beyond their room of manoeuvre. However, a troika member acknowledges that 'in the final days of the negotiations, it was very difficult to go back to the member states every time. In the endgame, we may have accepted things without a mandate of the EU coordination.' Moreover, member states found that the troika had given in on the targets and abandoned the -15% EU objective too early, even before the ministers had arrived in Kyoto.[18] This illustrates a tension between a

---

17   Member states in favour of article 175§2 TEC (stipulating unanimity) were Germany, Portugal, the Netherlands, Spain and Sweden. However, they could not modify the proposed legal basis by the Commission (article 175§1 TEC) (COM(2001) 579), as unanimity was required for this.

18   Certain member states took the view that the decision to deviate from the -15% objective could only be taken by the ministers, as this target was key in the Council

group of member states led by Germany, on the one hand, and the troika (mainly the UK at that time), on the other hand, about the strategy. The former wanted the troika to strictly maintain the -15% position, while the latter, experiencing the pressure of Japan and the US, opted to go down from that position earlier in order to guarantee progress at the international level. Moreover, the UK also seemed to have made commitments on sinks without the backing of the member states in a meeting with Japan and the chairman of the negotiations. As a consequence, it seems that the troika – and mainly the UK – strived to conquer additional autonomy vis-à-vis the member states, on top of the quite large degree of autonomy already granted by the member states.

4.4. Degree of preference homogeneity among principals <prefprin>

Although all EU member states had strong demanding positions compared to the external negotiation partners, their preferences were rather heterogeneous. On some issues (having quantified targets for emission reduction, including the EU bubble concept in the Protocol, stressing the importance of policies and measures) the principals' preferences did not diverge.

However, on other issues, 'member states were quite divided' (member state representative). First, member state preferences diverged on the height of the EU target, usually depending on their national situation. The negotiations on both the 1997 and the 1998 Burden Sharing Agreement were characterized as tough negotiations in the Council.[19] Member states like Spain, Portugal, Italy and Greece were opposed to very strong targets, while Austria, Sweden, Denmark and Germany were pushing for an overall target of -20%.[20] Second, member states were divided on how the quantified targets had to be reached. France, Denmark and Germany were initially very much opposed to the flexible mechanisms – mainly emission trading – and insisted on the need for policies and measures. Member states like Italy and the UK were more open to the emission trading system, as they considered it as a more pragmatic approach to achieve the targets. Third, principals like Sweden and Finland, two richly forested countries, were in favour of including sinks in the Protocol, while other member states opposed this for a long time.

---

conclusions, which were agreed upon at ministerial level.

19   Member states were not only divided on the different reduction percentages, but also on how the reductions should be calculated: in absolute values of tons of emissions (of which mainly Germany and the UK were in favour) or in terms of emissions per capita and taking into account the efforts a member state had already made to reduce its emissions (of which mainly France, Sweden, Spain and Portugal were in favour).

20   These four maximalist countries formed, together with Switzerland, the so-called 'Toronto Group', which strived to fulfil the targets of the 1988 Toronto Conference (i.e. all Annex I countries have to reduce their CO_2 emissions by 20% till 2005 in comparison with 1987 levels) (Jäger, O'Riordan, 1996). Austria, Sweden, Denmark and Germany, being the most maximalist member states, coordinated their views before EU meetings.

Consequently, the degree of preference homogeneity among the principals was not high. The cohesion member states were positioned at the minimalist end of the spectrum, while the Toronto Group, led by Germany and Denmark,[21] took the most maximalist preferences. The Netherlands and the UK had a somewhat more moderate – yet still maximalist – position, and France, initially being at the minimalist end of the spectrum, evolved to more maximalist preferences once the Green minister Voynet had taken the environment portfolio in the French government.

4.5. Degree of preference homogeneity between principals and agent <prefpa>

The Netherlands (being troika member in Kyoto and holding the Presidency in the first half of 1997) and the UK (being part of the troika) had rather maximalist preferences, but still somewhat more moderate ones than the Toronto Group. An official from a troika member state acknowledged that 'the members of the troika were indeed more on the maximalist side'. Moreover, the Netherlands seemed to enjoy a kind of ownership of the dossier, as they had a lot of renowned climate change experts, as they provided the EU member state with scientific input, and as they led the process in the first half of 1997, which led to a political deal on the Burden Sharing Agreement (Kanie, 2003).

The Dutch and the UK preferences were not always similar. As the Netherlands very much played the cards of policies and measures, while the UK was more in favour of flexible mechanisms to realize the emission reduction targets, it seems that these two views balanced each other in the troika. This was complemented with the Luxembourg Presidency, which did not have an outspoken national position.

4.6. Level of politicization <polit>

Climate change was – and increasingly is – definitely one of the most politically sensitive international environmental issues. Although the salience of the issue still has increased since 1997, the level of politicization was also high in the course of 1997. The Kyoto conference was characterized as one of the most intense international environmental negotiation sessions ever. The high degree of political sensitivity is illustrated by the large role of ministers in the negotiations, by the high degree of media attention, and by the political capital invested by a lot of (environment ministers of) EU member states in the issue (in particular Denmark, France, the Netherlands, Germany). The burden sharing was definitely a very important issue at the table of the Council of Ministers. It was politically sensitive for all member states, as it would influence their future policies in a lot of areas (energy, transport, industry, etc.). Moreover, certain member states had some

---

21   The Danish and German environment ministers, Svend Auken and Angela Merkel, were profiling on climate change issues. Auken, one of the most senior environment ministers in the EU, was considered as a strong personal advocate and leader in the process, while Merkel had been very active on the Berlin mandate (COP 1 to UNFCCC).

particular interests at stake in these negotiations, as the Nordic member states were e.g. very keen on the sinks issue.

4.7. Information asymmetry in favour of the agent <infag>

In the negotiations in the Subsidiary Bodies, the principals enjoyed the same information as the agent. However, in Kyoto – and certainly during the second week of the Kyoto meeting – the troika favoured a clear information benefit vis-à-vis the member states. This was certainly the case for negotiations between the US, Japan and the EU, and for the negotiations in a Friends of the Chair setting. A troika member described this situation as follows: 'It was so intense that it was very difficult to give everybody clear information. There were situations that we could not go back every time to the member states.' This is confirmed by a member state official, stating that 'there was obviously an inner circle [the troika] and an outer circle [the member states]. You have the feeling that you are not directly involved, that you only get second-hand information. The ministers were complaining at the end that they did not know what was happening.' Another member state representative affirmed that 'it was a hopeless task to follow the negotiations via the coordination meetings'. However, member states did not have the impression that the troika was strategically withholding information, and they admit that it was probably the intensity of the negotiations that caused this information asymmetry in favour of the agent.

4.8. Information asymmetry in favour of the principals <infprin>

The general impression among EU decision-makers with regard to the principals' information benefit is that the member states – including the Presidency and the troika members – knew each other's positions, but that they could not accurately assess what each other's fallback position was. It was particularly difficult for the EU negotiators to judge how much lower than the 15% reduction target they could go and which percentage was the minimal limit for the member states with most maximalist preferences. The principals did not put all their cards on the table with regard to the targets and the burden sharing. This was also the assessment of the EU negotiator, as a troika member declared: 'I am sure that some member states could have accepted more than they told in the EU coordinations.' Moreover, there are some individual examples of member states, which made clear during the process that they could not accept a provision, but nevertheless ultimately did: e.g., Germany mentioned that including sinks in the Kyoto provisions would be unacceptable to them, but it finally accepted.

4.9. Degree of compellingness of the external environment <extcomp>

The external environment generated a high degree of compellingness on the member states. 'We really felt a collective pressure' (member state official). As the EU was the most demanding party in the negotiations and as the ending Berlin Mandate produced a considerable time pressure in Kyoto, every member state wanted to avoid a failure of the negotiations. Moreover, in these negotiations, the cost of no agreement was particularly high because of the EU bubble concept. The fact that the member states needed an international agreement to guarantee a common reduction target, increased the pressure to agree on the final text and to ratify it afterwards. The compellingness of the external environment is illustrated by the following quote from a member state representative: 'At some point, the political costs were too high to stop it, even if we thought that we had not given the mandate to the Presidency for this or that'. Officials who were part of the troika acknowledge that they took into account that member states were willing to conclude a Protocol in Kyoto.

4.10. Institutional density <instdens>

The institutional density of the (first stage of the) EU decision-making process was rather high, although the EU as a whole can hardly be considered as one team in which no hard bargaining situations or conflicts took place. The Council's institutional norms played a role in the EU coordination meetings, although not to an extremely high extent. The cooperative spirit and team atmosphere between agent and principals certainly decreased in the final stages of the Kyoto negotiations, when there were some tensions on the involvement of the principals in the international negotiations. From the troika's perspective, these tensions were due to the fact that 'the EU negotiators knew where the negotiation partners had room of manoeuvre and that the member states could not understand this directly' (troika member official).

Diffuse reciprocity certainly played a role in the EU. Once the Burden Sharing Agreement of 1997 was agreed upon and once the EU bubble strategy was set up, it was clear to every member state that they were dependent on each other's realization of the reduction targets to fulfil their common international commitment of -8%. Moreover, the UK, as the future Presidency being a member of the troika, took into account the fact that it needed the principals during its upcoming Presidency period (January-June 1998). Indeed, the Burden Sharing Agreement had to be renegotiated after Kyoto into a legally binding agreement and this was one of the UK Presidency priorities.[22] Although understanding the

---

22   As the division of competences between the EC and the member states was not crystal clear on the burden sharing, there was a kind of gentlemen's agreement between the UK Presidency and the Commission that the Presidency could take the lead on the burden sharing during the first six months of 1998. However, if the Presidency did not succeed, the Commission would have taken over the dossier from July 1998 onwards. The UK wanted to avoid this at all costs, and this was an additional incentive for the UK to show some consideration for other the member states in Kyoto, in order to guarantee their cooperation in the months afterwards.

complete picture of each member state's situation in terms of ability to accept a particular emission reduction target was not possible for the member states, there was a general recognition and understanding of each other's problems. Such understanding, often caused by persuasive justifications of individual situations, seems to be necessary to understand the Burden Sharing Agreement, as this allows certain member states to increase their emissions, while other member states had to reduce them by as much as 30%. A member state official expressed this logic as follows: 'If there would not have been mutual understanding among the member states, how would you explain that France got 0% and Germany had to reduce 25%?' Research by Chagas, confirmed by another study by Vogler, shows that the notion of 'fairness' – and thus mutual responsiveness – was indeed decisive in the internal EU negotiations of the Burden Sharing Agreement (Chagas, 2003; Vogler, 2009).

Consensus was *de facto* the rule in the EU decision-making process with regard to the Kyoto Protocol negotiations. 'We always tried to have consensus and we never singled out a member state. When a country could not directly agree, it really came under pressure from other countries not to block the process' (official from a troika member state). The Luxembourg, UK and Dutch representatives expressed that they felt trusted by the member states, although member state officials acknowledge that Luxembourg was trusted more than the Netherlands and certainly more than the UK, because 'you trust more the impartiality of a small member state than the UK' (member state representative).

5. Aarhus Convention [ARHUS]

5.1. The international negotiating process

The Aarhus Convention grants rights to the citizens of its parties with regard to (1) access to environmental information, (2) participation in environmental decision-making, and (3) access to justice in environmental matters. These three packages of rights are the three so-called 'pillars' of the Convention (Palerm, 1999; Rodenhoff, 2002). The Aarhus Convention is the first international instrument carrying out Principle 10 of the 1992 Rio Declaration, which appeals to public participation in environmental decision-making. Unlike other MEAs, it does not stipulate rights and duties between the parties, but rights and duties between a party and its citizens (Veinla, Relve, 2005). As the Aarhus Convention was negotiated under the auspices of the United Nations Economic Commission for Europe (UNECE), it is a regional agreement.[23] The Convention was negotiated during ten one-week negotiation sessions between June 1996 and March 1998 and it was signed during the fourth ministerial 'Environment for Europe' conference in Aarhus (June 1998).

---

23   However, the Aarhus Convention is now open to every UN member state.

UNECE has 56 member states: mainly the countries on the Eurasian continent – including Russia and the other CIS (Commonwealth of Independent States) countries – plus Canada and the US. During the negotiations on the Aarhus Convention, the US and Canada only occasionally – and never actively – participated. As they did not sign the Convention, the parties to the Aarhus Convention are limited to European (although broader than EU) states and CIS countries (including Russia). The negotiation process was unique in the sense that it was characterized by a very active and influential participation by NGOs. They attended all negotiation meetings, intervened in the debates and proposed treaty articles.[24] A member state official described the role of the NGOs like this: 'They negotiated as if they were another country, quite a big country'. A Commission representative stated that 'they had an enormous impact on the negotiations'. The NGOs took the most maximalist positions at the international level, together with Norway and Poland. They strived for strong rights with regard to access to information, public participation and access to justice. Russia's position was located at the other end of the spectrum. In-between the NGOs and Russia, the EU member states took an intermediate position, although the member states' preferences significantly diverged (see further). Most Central and Eastern European countries had a position between the position of the NGOs and these of the EU member states. Several of these countries had domestic legislation on the Aarhus issues that was more progressive than the legislation in EU member states. However, as they mostly remained rather inactive in these negotiations, most of these Central and Eastern European countries were characterized as 'the silent majority' (member state official).

5.2. EU negotiation arrangement

The EU negotiation arrangement has evolved during the ten negotiation sessions. As a result, the EU decision-making process can be divided in two parts: the first seven sessions on the one hand, and the last three on the other hand (Delreux, 2009b). From session 1 to 7, there was no EU negotiator. In the few coordination meetings organized, the Presidency had to decide that there was no EU position. Consequently, the member states presented their own national positions in the international negotiations.[25] In that period, the Commission occasionally took the floor as well. In such situation, the Commission only spoke on its own behalf, not on behalf of the member states. Hence, the Commission did not act as agent in the first seven negotiation sessions.

This situation of no EU representation changed at the end of the negotiations, more in particular at the fourth day of the eighth negotiation session, when the

---

24   The only thing NGOs were not allowed to do, was attending EU coordination meetings. This was very contested by the NGOs.

25   Not all member states were closely involved in these negotiations. The most active member states were Germany, UK, Italy, Denmark, Belgium and the Netherlands.

mandate for the Commission was formalized. From then on, the Commission acted as EU negotiator for all the issues falling under EC competences. The first and the second pillar of the Aarhus Convention (on access to information and on public participation) were covered by Community competence because of existing EC legislation (on access to information: 90/313/EEC; on public participation: 85/337/EEC and 96/61/EC). The third pillar on access to justice remained under member state competences. As a consequence, the EU negotiation arrangement depended on the issue under discussion, resulting in a mixed system. On the one hand, the Commission was the agent for all the Aarhus Convention articles on which the EC was competent. However, even for these issues, member states could – and did – still take the floor. They sometimes supported the position as presented by the Commission, but they also sometimes ignored the EU position and expressed their own national preference, even if this was not in line with the coordinated EU position.[26] On the other hand, during the discussions on access to justice or on issues on which the member states had not succeeded to reach a common position,[27] they spoke freely in their capacity of UNECE member states, without being bound by EU discipline.

In the three last negotiation sessions, the Commission was only the agent in the plenary meetings. The various working groups, set up on an *ad hoc* basis, were always open for all representatives to fully participate, without EU representation. These working groups were characterized by (UNECE) member states experts coming together to solve a particular issue in a rather informal setting.

5.3. Discretion of the EU negotiator <disc>

The principals did not grant a large degree of autonomy to their agent. The Commission's negotiation authority was contested by the member states for a long time, i.e. until the eighth negotiation session. During three-quarters of the negotiation process, the principals did not delegate autonomy to any agent. Why did they then ultimately do so? All EU decision-makers admit that 'it was not about creating leverage' (member state official). Some of them mention a kind of blame-shifting reasoning by the member states when they authorized the Commission as their agent. In the first negotiation sessions, the NGOs had put pressure on the member states and they had influenced them to go in a more maximalist direction. As a result, in

---

26   An observer to the negotiations described such a situation as follows: 'If member states were following the statement of the Commission, they said something like "in support of the statement that has just been made on behalf of the European Community, I want to state that…" and then they said something that completely opposed to what the Commission had said.'

27   An example of an issue, which is related to access to information – and thus in principle EC competence – but on which the member states could not agree and spoke on their own behalf, was the provision on the confidentiality of industrial or commercial information (article 4§4/d of the Aarhus Convention).

these first negotiation sessions, several member state representatives went beyond what they could reasonably accept or what they were able to sell at their domestic level. As a consequence, member states ultimately granted negotiation authority to the Commission because they wanted to reconsider their initial positions to some extent, but simultaneously they also wanted to avoid losing face vis-à-vis the NGOs. A Commission official stated: 'In the beginning, some member states were going in the flow of the NGOs, but at the end, we had to do the bad job.'

As mentioned above, the mandate was only granted to the Commission after the seventh negotiation session (December 1997). However, the Commission proposal on the mandate had actually been sent to the Council in November 1996, i.e. between the second and the third negotiation sessions. The reasons for the delay of the mandate are threefold. First, the discussion on the mandate focussed on the Aarhus Convention's implications for the Community institutions. In the course of 1997, it was unclear whether the EC institutions would fall under the Convention's definitions of 'public authorities', which had implications for the openness of EU decision-making on environmental matters. Second, it was considered 'politically bad' (member state official) if the EU would have negotiated as a coordinated group in the UNECE, where there were *de facto* only Russia, Norway, the NGOs and a couple of Central and Eastern European states (mainly Poland and Hungary) playing a role. A member state representative described this as follows: 'We did not want to see the EU terrorizing UNECE, which was at the moment without the US and Canada.' As a result, the fact that pooled representation would increase the EU's political weight at the international level was not an incentive for delegation, as one would expect. On the contrary, it even restrained delegation. Third, the more maximalist member states tried to delay the mandate – and thus a common EU position – as they feared that a common position would be too minimalist and as they aimed to retain their own voice at the international level.

When ultimately agreed upon, the mandate only covered the issues on which there were EC competences (i.e. the first and the second pillars of the Aarhus Convention). Moreover, the mandate stipulated that the Convention could 'not be incompatible' with Community legislation. This formulation allowed that the option of a UNECE Convention going further than EC legislation was not excluded *a priori*. In other words, 'not being incompatible' only meant that that the existing EC directives 'were only the floor, not the ceiling' (observer to the negotiations). Because of this formulation, the maximalist member states agreed on granting negotiation authority and a mandate to the Commission.

In first negotiation sessions, member states and Commission usually held only one coordination meeting during the whole week.[28] These coordination meetings mostly 'ended up in chaos' (member state official), as no common position was reached. Once the Commission had the mandate and acted as EU negotiator for the first and

---

28   These coordination meetings were usually organized on Sunday afternoon, while the UNECE negotiation session started on Monday morning. During the third negotiation session, not one EU coordination meeting was organized.

the second pillars, the number of coordination meetings increased and they were held regularly (twice a day or more). Moreover, in the run-up to the two final negotiation sessions, member states twice met in Brussels to prepare these meetings. However, the coordination meetings still did not run smoothly. Even in the final negotiation session, a couple of coordination meetings ended without any internal EU agreement.

Member states were always able to participate in every international negotiation setting. Occasionally, the chair of the international negotiations called a Friends of the Chair setting, but this was not meant to bring the main players together and to force a decision. It was more a meeting of closely involved individuals who were there in their personal capacity to reflect on the continuation of the process. As a result, there were no negotiation settings in which only the agent and none of the principals participated.

During the negotiation stage, Germany threatened not to sign or to ratify the Convention at its domestic level.[29] Blocking the ratification at the EC level did not seem to be an issue. However, the ratification stage in the EU was not an easy process. It was not the ratification decision as such that generated problems, but the proposal of the implementation legislation. In October 2003, the Commission proposed a package of three legislative acts to implement the Aarhus Convention. The reason for presenting this as a package to the member states was that the Commission assessed the likelihood of reaching a qualified majority for the three acts separately as low. Hence, the Commission wanted to get it decided by the Council as a package deal. A first proposal concerned a regulation to apply the principles of the Convention to the institutions of the EC (COM(2003) 622). The second proposal was a directive intended to implement the third pillar of the Aarhus Convention,[30] as it dealt with access to justice for environmental matters in the member states (COM(2003) 624). The third proposal concerned the ratification decision by the EC (COM(2003) 625). When the Commission saw that its package deal strategy would not work (the opposition against the access to justice directive was too strong), it decoupled the proposals in order to get the Convention ratified before the second MOP to the Aarhus Convention (May 2005). That happened in time, as the Council adopted the ratification decision in February 2005 (2005/370/EC). The Commission considered EC participation in MOP 2 as crucial, because the MOP was going to negotiate and to adopt an amendment to the Convention, which would extend the rights of the public to participation in the decision-making on GMOs. The access to justice directive is still pending in first reading in the Council (dd. January 2011) and the regulation on the Community institutions was only adopted in September

---

29   Germany is the only EU member state that did not sign the Convention at the ministerial conference in Aarhus. It only signed the Aarhus Convention six months later.

30   By that time, the first and the second pillars of the Aarhus Convention were already fully implemented. The existing directives were adopted and updated to the Aarhus provisions (2003/4/EC; 2003/35/EC). These implementing acts were less controversial than the proposal to implement the access to justice pillar, as EC regulation already existed in these areas and as the existing directives foresaw a review by the end of the 1990s.

2006 (2006/1367/EC). At the national level, all member states, with the exception of Ireland, ratified the Convention, Germany being the last one in 2007.

All EU decision-makers admit that the Commission did not conquer additional autonomy. 'Once we were coordinated, the Commission was very careful,' according to a member state official. Thus, the agent did not go beyond the principals' instructions. Another member state official described the agent role of the Commission as follows: 'They were really representatives of the member states.'

5.4. Degree of preference homogeneity among principals <prefprin>

The degree of preference homogeneity among the principals was very low. Member states were divided on the Aarhus Convention in the sense that some did not want to go beyond the legislation that existed at the EU level, while others aimed to go much further. This preference heterogeneity caused a difficult EU decision-making process, frequently resulting in confusion and disagreement.

Germany's preference was considered as the most minimalist. The Aarhus issues fell under the competences of the German regions and they put the German representative under a strong pressure not to give in on provisions that would affect the transparency of their institutions. On the other side, Denmark,[31] Belgium and certainly Italy had the most maximalist preferences. They aimed to reach an international convention with stronger provisions – meaning provisions granting more rights to the public in the field of environmental democracy – comparing to what was already achieved at the EU level. Consequently, their preferences were situated closer to the NGO positions. The Netherlands and the UK were quite active member states holding rather moderate preferences. The UK even tended more to the minimalist side of the spectrum, as it was seen as 'not obstructing, but always very prudent' (member state official). The French preference was similar to the UK's (trying to avoid that the new Convention would have a large impact on its national legislation), but France was not an active player in the decision-making process.

5.5. Degree of preference homogeneity between principals and agent <prefpa>

As the Commission mostly took an intermediate position between Germany on the one hand and the maximalist member states on the other hand during the last three negotiation sessions,[32] the degree of preference homogeneity between principals and agent can be considered high. The Commission aimed to get an agreement on the first and second Aarhus pillars that did not diverge from existing EC legislation.

---

31 Denmark already had very transparent national legislation with regard to the Aarhus principles. Moreover, as Denmark would host the ministerial conference where the Convention would be signed, it was keen to reach an agreement that was quite strong, so that the NGOs would support it and not protest against it during the Aarhus conference.

32 As already mentioned, during sessions 1 to 7, the Commission did not play an active role in the negotiations.

Hence, its preferences were not as progressive as these of the maximalist countries (or the NGOs, at the international level).

This does not mean, however, that the Commission had no own institutional interest with regard to the Aarhus Convention. As the mandate stipulated that the EC had to become a party to the Aarhus Convention, its provisions would be applicable to the EC institutions, including the Commission. On access to information and on public information, the rules in the EU, mainly in the form of guidelines, were not only dealing with the environment policy area, but they were horizontal rules, covering all policy domains. So, the main interest of the Commission was to make sure that the Aarhus provisions would be consistent with these guidelines. The Commission's legal service emphasized that 'the Commission could not go unnecessarily far beyond what existed for the sake of the environment' (Commission official). This also explains the rather moderate preference of the Commission, in the sense that it was not a demanding party to adopt international rules that would affect Community institutions.[33] On access to justice, the EC Treaty was very restricted. Hence, on the third pillar also the Commission could not go very far.

5.6. Level of politicization <polit>

The issues of the Aarhus Convention were rather lowly politicized in the member states, probably with the exception of the UK, where domestic law on environmental democracy was being established at the same time. The fact that all negotiations took place at expert level is indicative for the low level of politicization: some decision-makers mentioned that there was no domestic coordination and no political instructions in a couple of member states. Moreover, the media attention for the negotiations was limited. Even the influential presence of the NGOs did not really increase the level of politicization. The strategy of the NGOs in order to make the Convention stronger was not to intensify the political sensitivity in each of the member states, but to influence directly the negotiations and the Convention text at the international level, which they were able to participate almost as if they were UNECE states.

5.7. Information asymmetry in favour of the agent <infag>

The Commission had no information advantage vis-à-vis the member states during the negotiations on the Aarhus Convention. First, the Commission played a very low-profile role during the first three quarters of the process. As a result, the member states were most actively involved in the negotiations. That did not change when

---

33 The EC made a declaration upon its signature (09551/98) and ratification (2005/370/EC) of the Aarhus Convention, stating that '[...] the Community also declares that the Community institutions will apply the Convention within the framework of their existing and future rules on access to documents and other relevant rules of Community law in the field covered by the Convention'. This illustrates the difficulties the EC had with the application of the Aarhus Convention to its institutions.

negotiation authority was delegated to the Commission for certain issues. Second, the principals could attend every negotiation session (in plenary and in working groups), as a result of which the Commission had no private information from restricted negotiation settings. Third, sharing the information it had with every principal was necessary for the Commission to keep all the member states on board in the difficult EU decision-making process, since the Commission strived to maintain a role in the process in order to defend the Community institutions' interests.

5.8. Information asymmetry in favour of the principals <infprin>

During the last negotiation sessions, i.e. when the Commission acted as EU negotiator, 'member states said what they wished' (member state official) in the EU coordination meetings. As a result, there was a rather low degree of information asymmetry in favour of the principals. The coordination meetings were the only forum where member states could be frank about their positions and their fallback positions, as it was the only setting to discuss issues with other member states without the presence of the NGOs.

5.9. Degree of compellingness of the external environment <extcomp>

For the Aarhus Convention negotiations, the external environment was somewhat – but not extremely – compelling. On the one hand, the fact that UNECE is a regional organization and that the EU had a relatively large weight in the negotiations because only Russia, some Central and Eastern European countries and Norway played a significant role, made the external environment not very compelling. EU decision-makers indicate that the European member states slightly dominated the negotiations. Hence, the pressure from other UNECE states on the EU to reach an agreement was quite low.

On the other hand, the prominent and influential presence of the NGOs increased the compellingness of the external environment, of which the NGOs were *de facto* totally part. As they publicly criticized the countries that tried to block a more progressive agreement, 'their presence increased the pressure', according to a member state official. Moreover, member states experienced a time pressure from the future ministerial conference in Aarhus, where the Convention was planned to be signed. It is generally acknowledged that the negotiations would have taken much longer and that there would probably not have been an agreement without the prospect of the Aarhus conference. This led to a 'pragmatic desire of having a result' and to 'a situation in which nobody wanted to be addressed as the one that made the negotiations fail' (member state officials).

5.10. Institutional density <instdens>

The EU decision-making process with regard to the Aarhus Convention was characterized by a very low degree of institutional density. The statements by member state officials that 'there was no European spirit' and that 'there were many tensions among the member states' indicate the low degree of institutional density in the EU.

Member state and Commission officials consider that diffuse reciprocity concerns did not play a role in the process: 'We did not take into account that there were other negotiation sessions following or that we would need each other for ratification or implementation. We even preferred to have no EU position than a common position that diverged from our owns.' Nor did the mutual responsiveness logic influence the EU decision-making process. None of the member states seems to have changed its preference because of the arguments or justifications presented by other principals. Germany e.g. did not manage to convince the other member states to take its internal concerns into account. During the EU coordination meetings, preparedness to compromise and to reach a consensus could hardly be discovered. If a member state did not agree with a majority position in the EU – even if it was about issues covered by EC competences – it took the floor at the international level to defend its own preference and it did not reconcile to the EU position. Finally, representatives from both the principal and the agent side indicated that the Commission was trusted by the member states as their negotiator. Trust was the only institutional norm that led the decision-making process to a somewhat more cooperative nature.

6. Rotterdam Convention on PIC [PIC]

6.1. The international negotiating process

The purpose of the Rotterdam Convention on Prior Informed Consent (PIC) is to protect human health and the environment against the unintended import of hazardous chemicals and pesticides. The PIC procedure is an information procedure on the basis of which parties are able to take informed decisions on the import of industrial chemicals and pesticides. Such a procedure on international trade in chemical substances already existed before the Rotterdam Convention came into force. In the 1980s, both FAO and UNEP developed a voluntary – and thus not legally binding – instrument promoting information exchange and shared responsibility between importing and exporting countries: the 'FAO Code of Conduct on Pesticides' and the 'UNEP London Guidelines on Chemicals' (Langlet, 2003). In 1995, the UNEP Governing Council decided that these voluntary procedures had to become legally binding. Therefore, it established an Intergovernmental Negotiating Committee (INC), which held five sessions between March 1996 and March 1998. These negotiations, characterized as rather technical, resulted in the Rotterdam PIC Convention.

In the beginning of the 1990s, the EU had already transposed the FAO Code of Conduct and the UNEP London Guidelines into an EC regulation (2455/92/EEC), which made the PIC procedure, which was at that time voluntary at the international level, mandatory for the EU member states. Because the EU did not want to remain the only one with a legally binding system and because it was convinced that an international treaty was needed to guarantee compliance, the EU took a strong and demanding position in the international negotiations (Bretherton, Vogler, 2000). The main divide in the negotiations was between a group of industrialized countries led by the US, striving for a weak Convention with a small scope, on the one side, and the EU on the other side (Kummer, 1999). Moreover, the US led group saw the Convention rather as a trade convention, while the EU considered it as an MEA.

6.2. EU negotiation arrangement

From EU side, the Rotterdam Convention was completely negotiated by the Commission (Delreux, 2008). As it concerned mixed negotiations, member states were formally allowed to take the floor, but they never did on substantive issues.[34] Hence, for negotiations in plenary, the main negotiating arena for PIC, the Commission was the only EU negotiator. The two main contact groups (the technical working group and the legal drafting group) were less formal and open for member states to take the floor. In the technical working group, most negotiating work was still done by the Commission. If the Presidency or other member states took the floor, they only did it to give additional arguments and to reinforce the EU position: even when member states theoretically got the chance to express their national preferences in certain negotiation settings, they never diverged from the agreed EU line. In the legal drafting group, the Commission's role was less prominent, as a couple, legal experts from the member states did a large part of the work. However, the legal drafting group can hardly be considered as a negotiating group, as this group's task was to put the texts coming from the technical experts into legal wording.

The strong role for the Commission – *de facto* as the sole EU negotiator, the only agent – can be explained by five reasons. The combination of these reasons made the PIC negotiations, as a Commission representative described it, 'a very special case, which will never happen again'. First, the PIC Convention largely deals with trade matters (this is also reflected in including article 133 TEC as the substantive legal basis of the Council decision on the signature and the ratification). As the Commission is the formal EU negotiator for trade issues, it was also allowed to play a large role in the PIC negotiations. Second, Regulation

---

34   When this occurred, it was e.g. the Dutch representative taking the floor to invite the conference to Rotterdam for the signature.

92/2455/EEC created the necessary competences for the Commission to act as EU negotiator. Even the issues that were not covered by the Regulation (compliance, financial issues, etc.) were *de facto* negotiated by the Commission.[35] A member state representative formulated it like this: 'The Commission had competences on PIC, but if certain issues went beyond its competences, they were entrusted to the Commission as well.' Member states showed a pragmatic approach in granting the Commission a prominent role in order to increase the EU's bargaining power, mainly vis-à-vis the US. Third, the PIC negotiations took place under the auspices of both UNEP and FAO due to the voluntary regulations existing under these two organizations. Contrary to its observer status in UNEP, the EC enjoys full membership status in the FAO. This strengthened the Commission's claim to lead the EU during these negotiations. Fourth, the experts from the member states came from their Designated National Authorities (DNAs), responsible for the implementation of Regulation 92/2455/EEC, and the Commission representatives were also in charge of the Common Designated Authority. As these member state representatives, who daily dealt with the chemical issues and who met very often in EU working groups and comitology committees to implement existing legislation, had seen the Commission acting properly in the implementation of Regulation 92/2455/EEC, they were confident that the Commission would do its job properly in the negotiations. Fifth, practical factors played a role as well. As INC 1, where the rules of procedure for the whole negotiation session were agreed, took place in Brussels, many Commission officials participated. Moreover, most third countries had sent negotiators from embassies in Brussels, who were familiar with the EU and who could more easily accept a role for the EC in the negotiations than representatives who would have come from the different capitals.

The exclusive agent role of the Commission can also be illustrated by the fact that EU coordination meetings before each INC did not take place in a Council setting (WPIEI) as it is the case for other negotiations, but in an *ad hoc* meeting in a Commission building.[36] These meetings were organized, prepared and chaired by the Commission. In-between the INCs, the Commission also negotiated bilaterally with third countries. For instance, between the third and the fourth INC, an important EU-US meeting was attended by a team of six Commission representatives and only one person from the Presidency.

6.3. Discretion of the EU negotiator <disc>

The Commission conducted the negotiations in close and good cooperation with the member states. The agent's negotiation authority was not contested by the principals: 'For the technical people, who were involved in these negotiations and

---

35   However, these issues were not very prominent in the negotiations: the financial and compliance mechanisms were mostly dealt with in the COPs afterwards.

36   Only the official authorization decision, together with the mandate, was taken in the Council.

who worked together with the Commission for other things, it was normal that the Commission would negotiate' (member state representative). It was taken for granted that the Commission, being the Common Designated Authority, having the competences on PIC and being trusted by the member states, would be authorized as EU negotiator. The Commission acted on the basis of a mandate, which referred to the existing secondary legislation. The Commission proposal for the mandate not only refered to EC competences on the basis of Regulation 92/2455/EEC, but also to the EC competences on the basis of the FAO Code of Conduct and the UNEP London Guidelines. This was unanimously rejected by the member states, which limited the competence to the internal PIC legislation. As a consequence, only the issues dealing with the PIC procedure as such – and e.g. not the financial issues – were covered by the mandate. However, this only seemed to be the formal story: during the negotiations, member states also allowed the Commission to negotiate on their behalf on those issues that were not covered by Regulation 92/2445/EEC *strictu senso*. This 'additional authorization' was granted in the regularly organized coordination meetings. These meetings generated clear EU positions, always based on the existing internal legislation. In that sense, the member states granted a high degree of autonomy to the Commission, although the Commission was not allowed to negotiate a Convention that went beyond EC legislation.

Member states were always present during the international negotiations (both in plenary and in contact groups), sometimes giving advice to the Commission on the spot. However, when the chair of the international negotiations organized informal consultation meetings with the main players, not all EU member states participated. But, nor was only the Commission representing the EU. In these meetings, which did not take place very frequently, a group of closely involved and interested member states accompanied the Commission.[37] That group of member states naturally grew out of the process and had more frequent and strong contacts with the Commission in the preparation of each INC.[38]

In the negotiating stage, there were no real threats of non-ratification by the EC or a member state. However, in the final hours of INC 5, when everything seemed to be settled, a Danish representative, who had only been involved in the negotiations during the last days, insisted that the EU should make a declaration on the relation between the PIC Convention and other international agreements.

---

37   For the PIC negotiations, the more closely involved member states were Germany, the Netherlands, France, Sweden, Austria, Belgium and UK.

38   The Commission e.g. invited these key member states for consultation on technical issues (not on financial or compliance issues) in Brussels. Neither representatives from the member states nor Commission officials considered these consultations as 'something mysterious' (member states representative). Representatives who attended such informal meetings describe them as 'an exchange of ideas' or 'giving support to the Commission'. Member states that did not play an active role, did not take offence at these meetings. Everything that was prepared in these meetings was adopted in a general coordination meeting afterwards.

This issue, on which Canada and the US had very outspoken positions, had been negotiated over many days, and at the final moment Denmark almost blew it up. At that moment, the Commission faced a real involuntary defection risk. The issue was solved by the Commission calling the Danish minister, telling him 'if you insist on this, we have to call a Coreper meeting now and have you outvoted' (Commission representative). In the ratification stage in the Council, there was no discussion on the Convention's substance or on the question whether the EC should ratify it. However, the legal basis of the Community's ratification decision became an issue under discussion. The Commission's ratification proposal (COM(2001) 802) stated articles 133 and 300 TEC as the legal basis. However, supported by the opinion of the European Parliament (A5-0290/2002), the Council unanimously rejected trade article 133 TEC and replaced it by the environment article 175§1 TEC (2003/106/EC). The conflict between Commission (wanting article 133 TEC, reasoning that the main objective of the PIC Convention relates to rules and procedures for international trade of certain hazardous chemicals) and Council (taking the opposite view that the Convention primarily aims at protecting the environment, and therefore wanting article 175§1 TEC) was ultimately solved by the judgement of the ECJ in case C-94/03. The Court ruled that the ratification decision should be based on the dual substantive legal basis of article 133 and article 175§1 TEC. As a consequence, the ECJ annulled the ratification decision that was taken on the basis of article 175§1 TEC and the Council adopted a new ratification decision, now based on the dual legal basis of articles 133 and 175§1 TEC *juncto* article 300 TEC (2006/730/EC). This procedure had neither consequences for the internal EU decision-making process (voting rules, role for the EP, etc.) nor for the international consent to be bound by the EC. This difficult ratification process, characterized by a conflict between agent and principals, can nevertheless not be considered as a sanctioning of the Commission by the member states because of the former's unfair negotiation behaviour in the negotiation stage or because of a dissatisfaction by the latter about the PIC Convention. The Convention was also ratified by each member state separately between 2000 and 2005.

The relation between the agent and the principals was considered very good, cooperative and supportive. As a Commission official noted: 'We did consultations all the time. The member states were close to us and we were discussing directly with them in the plenary.' This was confirmed by the member states. In the negotiations on the PIC Convention, the EU negotiator did not attempt to conquer more autonomy than the autonomy granted by the member states. There was never a question of the Commission going beyond the instructions. A member state representative expressed it as follows: 'I don't recall any moment in the negotiations in which the Commission has said something that was not in accordance with the common member states position. The Commission has negotiated in a very trustworthy and professional way. It was a very good cooperation with the Commission.' This is also confirmed from the Commission side: 'We knew how far we could go, because we had our Regulation as a model. It was clear for us where we wanted to go and how far the member states could go.' Commission representatives also stressed the fact that the Commission being the sole EU negotiator in international environmental negotiations was new and unprecedented. They aimed to avoid making a bad impression in this situation where they enjoyed a high degree of granted autonomy. Therefore, 'not to put in danger this whole experience', as stated by a Commission official, was an incentive for the Commission not to behave opportunistically, e.g. by conquering more and more autonomy. From this perspective, the Commission anticipated on a next delegation decision by the member states and it wanted to avoid being confronted with principals penalizing the Commission in the future by not granting such a high degree of autonomy anymore.

6.4. Degree of preference homogeneity among principals <prefprin>

Member states' preferences on PIC were very homogeneous. This can be explained by a twofold reason. First, the PIC negotiations dealt with a procedure that was already in force in the EU. The joint position of the member states was to make the procedure, which already existed in a legally binding way in the EU and on a voluntary basis in the international system, internationally legally binding. Second, as all member state representatives came from their DNAs, their common interest was that the rules, for the implementation of which they were responsible, were followed. Consequently, EU member states jointly took a quite maximalist position. Within the small degree of variation in member states' preferences, the UK's position was somewhat less maximalist, whereas the Netherlands and the Nordic countries (Finland, Sweden and Denmark) had the most maximalist positions. However, the preference distance between these two extremes in the EU was very small.

6.5. Degree of preference homogeneity between principals and agent <prefpa>

In the negotiation stage, there was no significant preference distance between the member states on the one hand and the Commission on the other hand.[39] As a consequence, the EU as a whole, consisting of an agent and 15 principals, had homogeneous preferences. Like the member states, the Commission wanted to reach a convention that was as strong as possible and that remained within the limits of the EU legislation. Given this situation, there was no question – or risk, from the perspective of the member states – that the Commission would use its role as EU negotiator to promote or to realize its own preferences.

---

39   In the ratification stage, there were more tensions between agent and principals, because of the discussion on the legal basis of the ratification decision.

6.6. Level of politicization <polit>

The PIC negotiations were not politically sensitive in the member states. They were mainly conducted by technical experts. There were no ministerial segments in the INCs, and the media did not pay attention to the negotiations. In the 1980s, i.e. before and during the discussions on the UNEP London Guidelines and the FAO Code of Conduct, the PIC procedure was more politicized. Yet during the INCs of 1996-1998, the PIC procedure was already politically accepted, certainly in the EU member states. Moreover, there was very little at stake in the EU to be politically sensitive: the chemicals under discussion were already regulated by Regulation 92/2455/EEC.

6.7. Information asymmetry in favour of the agent <infag>

Generally speaking, during the INCs, member states could have the same information as the Commission had. They attended all plenary meetings and they were able to be present in the contact and working groups, which had an open-ended participation. In practice, member states – with the exception of the less interested countries like Spain, Italy, Greece and Ireland – mostly participated in these contact groups. In none of the negotiation settings did only the Commission represent the EU without the presence of a member state. As a consequence, member states could easily have the information on the substance and developments on the international negotiations. In addition, member state representatives acknowledge that the Commission never strategically played with information and that the Commission provided the member states with information whenever they asked for it.

However, in-between the INCs, the Commission sometimes met informally with some external negotiation partners. For such bilateral meetings, the Commission invited the most interested and most closely involved member states – also because these member states had the best technical knowledge on PIC – to accompany them. Hence, only in such meetings, some member states did not have all the information.

Consequently, the Commission certainly possessed some information that some less interested member states did not have. However, the really involved principals could have all the information. As an official from an active member state said: 'From a member state with an active position, I never had the impression that the Commission had more information than we had.'

6.8. Information asymmetry in favour of the principals <infprin>

Like the Commission shared its information about the international negotiations collegially with the member states, the member states did so vis-à-vis the Commission regarding their preferences, fallback positions and the range of possible agreements they would ultimately be able to accept. The impression of the Commission that 'member states were playing quite openly' was confirmed by the principals: 'It was not the case that member states had hidden agendas or fallback positions vis-à-vis the Commission.'

6.9. Degree of compellingness of the external environment <extcomp>

As the EU had the most demanding positions in the international negotiations, it experienced the pressure from other countries and negotiation blocks, mainly led by the US, that wanted a less demanding agreement. There was not only a pressure on the EU, but also on the member states separately. A member state official expressed this as follows: 'We all wanted an agreement.'

6.10. Institutional density <instdens>

The EU decision-making process with regard to the negotiations leading to the PIC Convention was characterized by a rather dense institutional environment. Both Commission and member state representatives confirm that EU negotiations took place 'in a very good atmosphere'. The fact that the people involved in the negotiations were experts, coming from DNAs and having negotiated with each other and with the Commission representatives long before the issue of an international treaty on PIC was on the table made that mutual responsiveness and diffuse reciprocity influenced the negotiation behaviour in the EU decision-making process: 'Member states were trying to understand each other's domestic problems. They were not ignoring the positions of the others, they were taking it into account' (Commission official). A member state official even acknowledged that 'some member state experts were influenced by the input of other member state experts'. Moreover, member states were aware of the fact that they would need their negotiator, the Commission, in the future, as the Commission would take a central role in the notification procedure built into the PIC Convention. The institutional environment was characterized by a high degree of trust as well, both among the member states and between the member states and the Commission. A Commission official noted: 'We felt trusted by the member states. And vice-versa: we trusted the member states too'. This was confirmed by the member states representatives: 'People were very trusting each other. The people in the coordination meetings were people the Commission knew for years. They really had trust in these people. And we trusted the Commission. In fact, it was more trust in persons than in the institution.' Moreover, the Commission and the involved member states had frequent and regular e-mail and phone contacts in-between the different INCs, which was beneficial in building one EU team. Finally, there was a continuous striving for a consensus in the EU. As the general EU line to strive for a conversion of the EC Regulation to the international level was not contested by any member state, a consensus was always found relatively easily. The fact that the consensus was ultimately threatened by a Danish representative who was new in the process (see above) shows that the dense institutional environment had generated a socializing and consensus striving power among the decision-makers who were involved during the whole EU decision-making process.

7. Cartagena Protocol on Biosafety [CART]

7.1. The international negotiating process

The Cartagena Protocol on Biosafety is concluded under the 1992 Convention on Biological Diversity (CBD), one of the so-called Rio conventions, *inter alia* setting principles for the conservation and sustainable use of biological diversity. The Cartagena Protocol aims to protect biodiversity by regulating the safe transfer, handling and use of genetically modified organisms (GMOs)[40] that may have adverse effects on the conservation and sustainable use of biodiversity (Thieme, 2001; Kirsop, 2002). More specifically, the Cartagena Protocol establishes an internationally legally binding framework for the transboundary movement of GMOs (including international trade). The Protocol stipulates a procedure by which GMO exporters have to obtain an Advance Informed Agreement (AIA) (i.e. an explicit consent based on a risk assessment) from the importer before certain GMOs can be exported (Falkner, 2000; Andrée, 2005).[41] Moreover, the Cartagena Protocol permits the restriction of GMO import (i.e. not granting a consent) on the basis of precaution, as the precautionary principle is one of the guiding principles for risk assessments (Kirsop, 2002; Bevilacqua, 2007).[42] Once an importer has decided to accept the GMOs, the exporter has to provide the exported GMOs with appropriate labelling and documentation requirements.

CBD's COP 2 in Jakarta (1995) established an Open-ended Ad Hoc Working Group on Biosafety (the so-called BSWG, or Biosafety Working Group) to negotiate a Protocol regulating biodiversity-related GMO issues. The BSWG held six negotiation sessions between July 1997 and February 1999. The sixth BSWG was held in Cartagena and was meant as final preparation for the Extraordinary COP (ExCOP) of CBD that would take place immediately afterwards in order to conclude an agreement of the Protocol. However, the ExCOP collapsed, mainly because the US could accept neither the chair's compromise text nor a final

---

40 The terminology used in the Cartagena Protocol is not 'GMOs', but 'LMOs' (living modified organisms). The scope of the Protocol does not include GMOs for pharmaceutical use.

41 More precisely, the Protocol specifies two different procedures: on the one hand for GMOs that will be introduced in the environment (the AIA procedure), and on the other hand for GMOs that are meant for food, feed or processing (the so-called 'commodities') (an information procedure) (Burgiel, 2002).

42 The precautionary principle is often defined as follows: 'When an activity raises threats of harm to human health or the environment, precautionary measures should be taken even if some cause and effect relationships are not established scientifically' (Raffensperger and Tickner, 1999, p. 8).

compromise proposed by the EU (Bail, Decaestecker, Jørgensen, 2002; Rhinard, Kaeding, 2006).[43] At 24 January 1999, the Chair of the ExCOP could only conclude that the major negotiation partners were not able to find an agreement. As a result, the ExCOP was suspended. After a six-month cooling-down period, informal consultative meetings were held in Vienna, where the main players confirmed their political will to reach an international agreement on biosafety. In that period, GMOs and biosafety were at the top of the international political agenda (Bail, Decaestecker, Jørgensen, 2002). In January 2000, negotiation parties met again in Montreal to continue the ExCOP (ExCOP-bis). After some days of tough negotiations, all parties finally accepted a compromise text in Montreal (Winfield, 2000).

The EU took an intermediate position between the majority of developing countries on the maximalist side and the GMO exporting countries on the minimalist side (Bretherton, Vogler, 2000). From BSWG 6 in Cartagena to the ExCOP in Montreal, which was the period of the most intense negotiations, the negotiations were conducted by five groups of countries with similar interests (Gupta, 2000; Burgiel, 2002; Falkner, 2002). The Miami group, composed of countries exporting agricultural products and concerned about the trade implications of a biosafety agreement, had the most minimalist positions.[44] The Like-Minded Group (i.e. the G77 without the agricultural exporters Argentina, Chile and Uruguay) was positioned at the other side of the spectrum. The EU occupied a position in the middle, like it was the case for the Central and Eastern European Group (including Russia) and the Compromise Group, which assembled the non-EU and non-agricultural exporting OECD (Organization for Economic Cooperation and Development) countries.[45]

Although the EU occupied an intermediate position at the international level, it had some clear preferences about the Protocol: opposing a 'savings clause', which would subordinate the Protocol to WTO agreements, striving for a broad scope so that a lot of GMO categories would be regulated, attempting to include an AIA procedure, labelling and documentation requirements and a detailed methodology for risk assessment (Bail, Decaestecker, Jørgensen, 2002). Moreover, the EU became a strong advocate of the inclusion of the precautionary principle in the Protocol in the course of 1999 (Graff, 2002; Falkner, 2007). While the EU accepted

---

43   This EU compromise did not include an explicit reference to the precautionary principle, which was nonetheless considered as very important by the EU (and which was opposed by the US). This shows that the EU really wanted to go for an agreement in Cartagena. However, the US did not accept it. Ironically, the final agreement on the Cartagena Protocol, accepted by the US, contains four explicit provisions on precaution.

44   The members of the Miami group were Argentina, Australia, Canada, Chile, Uruguay and the US. None of these countries has ratified the Cartagena Protocol (dd. January 2011).

45   The members of the Compromise Group were Japan, Norway, South Korea, Mexico, Switzerland, Singapore and New Zealand.

a compromise that did not explicitly mention precaution in Cartagena, it would not have done so in Montreal a year later.

7.2. EU negotiation arrangement

The EU negotiation arrangement and the actors taking the agent role evolved during the course of the negotiations. In the context of CBD, it is normally the Presidency negotiating on behalf of the EU. However, at CBD's COP 2, which resulted in the Jakarta Mandate establishing the negotiations leading to the Cartagena Protocol, the Commission challenged the Presidency's role because the Jakarta Mandate touched upon trade issues as well (Thieme, 2001). During BSWG 1, there was an institutional conflict about who should speak on behalf of the EU. The Commission had convinced the Irish Presidency that it had to be the sole EU negotiator, arguing that all issues were falling under Community competences. By contrast, the member states wanted the Presidency to continue to do its job. They were concerned that the Commission did not have enough expertise on the technical aspects of biotechnology to negotiate the whole range of issues under consideration. However, the Irish Presidency was apparently not able to prevent the Commission from negotiating all the issues, and the Commission was thus *de facto* the only agent in BSWG 1.

As a reaction to this tension between the Commission and the member states, the Netherlands – being the Presidency during BSWG 2 and, according to a Commission official, being 'very much behind the Irish Presidency in trying to prevent the Commission of negotiating alone' – challenged the EU negotiating arrangement of BSWG 1, arguing that biosafety is a matter of shared competence. To solve this problem, the Commission and the Presidency negotiated a detailed list with the main topics of the future Protocol and who would speak on that topic. This gentlemen's agreement between the Commission and the Presidency established a division of labour, which was used during BSWG 2-6. From then on, for every item that popped up at the international negotiation table it was clear whether the Commission or the Presidency would be the EU negotiator. All trade-related issues and issues with a connection to the already existing GMO legislation in the EC[46] (including AIA, definitions, etc.) were handled by the Commission, while the Presidency was the agent for more technical issues (exchange of information, liability, capacity building, etc.).[47]

During the BSWGs, negotiations were held in plenary, working group and contact group settings, all of which could be attended by the principals. However, at the ExCOP in Cartagena, this negotiation setting changed because of a twofold

---

46   EC competences on GMOs originate from the Directive on the contained use of genetically modified micro-organisms (90/219/EEC) on the one hand and the Directive on the deliberate release into the environment of GMOs (90/220/EEC) on the one hand.

47   The complete division of labour between Commission and Presidency can be found in Thieme (2001), p. 263.

reason. First, interest-based negotiation groups were established (see above). Second, the ExCOP chair organized a kind of Friends of the Chair meetings, which were restricted in nature and composed of representatives of these interest groups, complemented with the co-chairs of some of the working and contact groups. In Cartagena, this negotiation setting was called the 'Group of Ten'.[48] The Commission occupied the only EU seat at the table and was the official EU negotiator in the Group of Ten. One representative of the Presidency could assist the Commission in this setting. However, as a government change in Germany had just taken place before the Cartagena meeting, the German Presidency team was internally divided and could not take a strong role. It was said that 'the Germans practically left the negotiations to the Commission' (Commission official). As a consequence, the Commission *de facto* became the only EU negotiator in Cartagena (Bail, Decaestecker, Jørgensen, 2002).

After the negotiations collapsed in Cartagena, the process continued with informal consultation meetings, which started in the summer of 1999, mainly in Vienna. These informal talks took place in a new negotiation setting: the Vienna Setting. In this setting, each group had two seats at the negotiation table. Other countries could attend the meetings as well by sitting behind their representatives, but they could not intervene. By organizing the negotiations like this, transparency – and with it the likelihood of reaching an agreement – increased (Gupta, 2000; Falkner, 2002). The Vienna Setting was not only used in the consultations meetings in the second half of 1999, but also in the final negotiation session in Montreal (ExCOP-bis). The Commission and the Presidency were the two EU representatives in this setting. As a result, the agent on each topic was present, and Commission and Presidency representatives could consult with each other during these very intense meetings. They were assisted and supported by member state experts sitting behind them. In Montreal, the EU decision-making process got an additional dimension, as ten environment ministers from the member states and Environment Commissioner Wallström attended the ExCOP-bis. The ministers did not take over the negotiating role from the civil servants,[49] but 'they played a decisive role for the final agreement while staying away from the formal process' (Bail, Decaestecker, Jørgensen, 2002, p. 182).[50] Table 4.1 summarizes the different negotiation settings in which the Cartagena Protocol was negotiated.

---

48   The Group of Ten was composed of the chair of the ExCOP, five representatives of the Like-Minded Group, two representatives of the Miami Group and one representative of the EU, the Central and Eastern Europe Group and the Compromise Group (Bail, Decaestecker, Jørgensen, 2002; Falkner, 2002).

49   This was mainly because not every external negotiation partner was represented at ministerial level. Sending the ministers to the international negotiations would have created an imbalance at the international level.

50   Ministers had bilateral meetings with the Miami Group, they organized press conferences, etc. In sum, 'they did a lot at the political level to save the Protocol and to back the negotiators and the officials' (Commission representative). A member state representative

Table 4.1   Negotiation settings and EU negotiation arrangement during the Cartagena Protocol negotiations

Negotiation session	Place	Negotiation setting	Agent	Role of principals
BSWG 1 (07/1996)	Aarhus		Commission	could attend
BSWG 2 (05/1997)	Montreal			
BSWG 3 (10/1997)	Montreal			
BSWG 4 (02/1998)	Montreal		Commission + Presidency	could attend
BSWG 5 (08/1998)	Montreal			
BSWG 6 (02/1999)	Cartagena			
ExCOP (02/1999)	Cartagena	Group of Ten	Commission	could not attend
Informal consultations	Vienna	Vienna Setting	Commission + Presidency	could attend
ExCOP-bis (01/2000)	Montreal			

During the whole negotiation process, the EU was exclusively represented by the Commission or the Presidency. Member states did not take the floor at the international level. Occasionally, a member state expert was invited by the EU negotiator to explain a technical issue. As this always occurred under the umbrella of the EU negotiator, the expert never represented (the preferences of) his or her member state. The situation in the various contact groups was from time to time somewhat different, as sometimes the EU position was expressed by a member state. These were experienced experts, who were unofficially chosen by the EU coordination meeting to represent the EU in a contact or working group. However, the position expressed by such a member state was always on behalf of the EU. The philosophy behind this preliminary lead country approach was that 'we had individuals in the EU who were very good in a particular issue. We used them and we put them in when they were needed' (member state official).

Probably the most important observation about the EU negotiation arrangement during the Cartagena Protocol negotiations is the evolution from a rather conflictual

declared: 'We needed the ministers because we needed the political responsibility to agree on the final text.'

situation in Jakarta and at BSWG 1 to a dense and well-functioning team at the end of the negotiations. The EU evolved from a formulaic to a pragmatic way of cooperation.[51] According to a member state official, 'it was not a set of arrangements as you read in the Treaty. The point was that we had people. Over the years, there was a great deal of trust and respect between individuals. The EU became a homogeneous body, where individuals took on various tasks. There was a real growing together, which was the real strength of the EU. It was a sense of joint ownership and joint purpose where everybody wanted to play a role in.' A result of this growing team spirit in the EU was that the gentlemen's agreement, stipulating the division of labour between Commission and Presidency, was no longer strictly used in the final stages of the negotiations. A member state representative stated it like this: 'The formal division of labour was quietly put aside in favour of getting the best people in there.'

7.3. Discretion of the EU negotiator <disc>

Only at CBD's COP 2 in Jakarta and at BSWG 1 in Aarhus, was the negotiation authority of the Commission contested by the member states (Bail, Decacstecker, Jørgensen, 2002). After these tensions were solved by the gentlemen's agreement between Commission and Presidency, the Commission's role as agent was no longer challenged. Neither was the role of the Presidency in representing the EU ever questioned. 'It was very clear from the beginning that we would go into these negotiations with an EU process' (member state official).

The Commission negotiated under a formal mandate by the Council. The mandate as such was quite broad, as it did not include strict substantive instructions. However, the mandate was complemented by Council conclusions, which elaborated in a more detailed way the common member state positions. During the negotiation process, there were three versions of the Council conclusions. First, the Council conclusions of October 1995 (10208/95) are considered as 'vague, in the sense of "wait and see"' (Rhinard, Kaeding, 2006). They were issued at the time of the COP 2 in Jakarta, when the EU position could still be considered as a two-track approach, as a legally binding protocol was only seen as one option to deal with the biosafety issue (Falkner, 2007). Second, the Council conclusions of June 1996 (8518/96) were adopted to make the negotiation mandate appropriate to the Jakarta Mandate, which set the general framework for the Cartagena Protocol negotiations. Third, the December 1999 Council conclusions (13854/99), adopted just before the final Montreal meeting, stressed the importance of reaching an agreement on the Protocol and allowed more leeway for the EU negotiator, although it was strict on what the EU negotiator could accept with regard to the

---

51   Bail, Decaestecker and Jørgensen describe this as follows: 'All member states knew that there were crucial decisions to be made at Cartagena and that this would not be the time for internal EU wrangling over formalities' (Bail, Decaestecker, Jørgensen, 2002, pp. 174-175).

precautionary principle and to the Protocol's relation to WTO agreements (Bail, Decaestecker, Jørgensen, 2002; Rhinard, Kaeding, 2006).

The Council conclusions served as the basis for the EU position papers that were prepared in Brussels before each international negotiation session[52] and specified during on the spot coordination meetings, which took place frequently. For every negotiation session – probably with the exception of the Cartagena and Montreal meetings – the Presidency and the Commission, in close cooperation with the principals, elaborated detailed and outlined position papers, mostly including fallback positions and possible room for manoeuvre. The principals' possibility to attend the international meetings has already been mentioned above: in principle, member states were able to participate in every negotiation setting, with the exception of the Group of Ten in Cartagena.

None of the member states or Commission officials recalls a situation where a principal threatened not to ratify the Protocol or to jeopardize the negotiations. Consequently, the EU negotiator was not confronted with a possible involuntary defection risk. The EC ratification is considered as a formality: 'There were no political difficulties' (member state official). Another member state representative stated: 'We were not happy with all the results, but of course we could live with it'.[53] Before the Commission initiated the ratification decision, Commission and Council had a different opinion about the appropriate substantive legal basis for the ratification decision (van Calster, Lee, 2002). While the Commission opted for articles 133 and 174§4 TEC *juncto* article 300 TEC, the Council's position was to base the decision on article 175§1 TEC *juncto* article 300 TEC. On the one hand, the Commission reasoned that the Cartagena Protocol has a trade-based content and that therefore article 133 TEC (common commercial policy) should be the substantive legal basis. However, the Commission accepted that the Protocol deals with environmental protection and opted for article 174§4 TEC as the substantive legal basis. Consequently, the Commission argued that the member states only retained concurrent powers for the issues that do not affect trade in GMOs. On the other hand, the Council's position was that the Cartagena Protocol is essentially an environmental agreement, regulating biodiversity. In its Opinion 2/00, the ECJ stated that, since the Cartagena Protocol is 'an instrument intended essentially to improve biosafety and not to promote, facilitate or govern trade' (Opinion 2/00, paragraph 36), article 175§1 TEC *juncto* article 300 TEC is the appropriate legal basis. Following this Opinion, the Commission based its ratification proposal on article 175§1 TEC *juncto* article 300 TEC (COM(2002) 62). The Cartagena

---

52   The coordination meetings in Brussels took place in the so-called 'Ad Hoc Group on Biosafety', being a subgroup of the Council Working Group on the Environment.

53   For example, some member states had difficulties with the flexibility in the Protocol on the labelling requirement for GMO transports. That was the last compromise made by the EU to get the agreement accepted by the US in Montreal (Bail, Decaestecker, Jørgensen, 2002; Falkner, 2002).

Protocol was finally ratified by the Council in 2002 (2002/628/EC). Between 2002 and 2004, every member state ratified the Protocol at its domestic level.

The EU negotiators did not go significantly beyond the principals' instructions. However, a member state official recalled that 'the Commission or the Presidency sometimes went too fast', and another member state representative stated: 'I sometimes had the impression "we have not agreed it like this", but that was never on important issues, it was rather to make progress'. Moreover, at the end of the Cartagena and Montreal meetings, the EU negotiators could not come back to the coordination meeting every time before they agreed on an issue. All member states acknowledge, nonetheless, that they were never confronted by a *fait accompli* of great consequence by the Presidency or only the Commission. Member state officials confirm that 'our negotiators, from the Commission or from the Presidencies, always dealt very correctly with the issue'.

7.4. Degree of preference homogeneity among principals <prefprin>

During the first two years of the negotiations, the degree of preference homogeneity among the principals was low. Germany and France – and to a lesser extent also the UK and the Netherlands – had minimalist preferences (Rhinard, Kaeding, 2006).[54] They initially questioned the need for a Protocol as such. Once the decision to establish a legally binding agreement was taken, these minimalist member states were sceptical to the inclusion of heavy procedures in the Protocol (e.g. a strong AIA procedure). To put it frankly, these member states were in favour of biotechnology and opposed to a strong biotechnology-restricting Protocol. On the other side, Sweden, Denmark and Austria took maximalist preferences, striving for a strong, GMO-trade-restricting Protocol (Bail, Decaestecker, Jørgensen, 2002).

When the negotiations evolved, the degree of preference homogeneity among the member states increased. The switch from heterogeneous to homogeneous preferences took place in the course of 1998 (i.e. around BSWG 5) and it was caused by the initially minimalist member states (mainly France and Germany) changing their preferences in a more maximalist direction. This can be explained by two reasons. First, all issues relating to GMOs had became much more sensitive in Europe. The debate on biosafety, in which the general public opinion went in the direction of anti-biotechnology, and the first GMO put on the market in Europe increased the sensitivity of the issue (Falkner, 2007). Second, a government change took place in Germany, where a red-green coalition had replaced a conservative one. Also in France (with the entrance of a green environment minister in government)

---

54   As the main German concern was the pharmaceutical industry, Germany wanted pharmaceutical GMOs to be excluded from the Protocol's scope. France was the largest agricultural exporter in the EU and was a potential exporter of GMOs. That explains why France was hesitant in the beginning and why its preference was more on that side of the preference spectrum that went in the direction of the Miami Group (the main agricultural exporters).

and in the UK (where the Labour government took over from the Conservatives), political parties that are traditionally more opposed to biotechnology came into power. As a result, the principals' preferences converged in a more maximalist direction. This strengthened the internal EU cohesion and led to a 'more unified and a more harmonized approach of the EU as a whole' (Commission official).

7.5. Degree of preference homogeneity between principals and agent <prefpa>

The degree of preference homogeneity between principals and agents was high during the whole process. Both the Commission and the various Presidencies did not have preferences that diverged from the common member states' position. The Commission's preferences were moderate and located in the middle of the range of member state preferences. Although the Commission was initially seen as rather minimalist by the maximalist member states like Austria and Sweden, this was mainly because the Commission insisted that nothing in the Protocol could contradict existing EC legislation on GMOs.

Also the other agents – the various Presidencies – did not have diverging preferences. During BSWG 1 to 5, Ireland and Luxembourg, respectively holding the Presidency during the first and the third BSWG, did not have very outspoken national positions.[55] The Netherlands and the UK (in BSWG 2 and 4) had preferences somewhat more on the minimalist side, while Austria (in BSWG 5) can be situated on the maximalist side. However, even during this period of rather heterogeneous principals' preferences, 'none of these Presidencies had national positions that were different from the EU position' (Commission official). As a consequence, the degree of preference homogeneity between the principals and the negotiating Presidencies can be considered high. Germany held the Presidency during the Cartagena meeting, but due to the recent government change, the internal conflict in the German delegation was bigger than the conflicts in the EU. Hence, the member state Germany did not have a real preference at the time of their Presidency. The same goes for Portugal, holding the Presidency during the Montreal meeting, and little interested in the issue. Moreover, as stated above, the agent role was *de facto* quasi-completely played by the Commission at that time.

7.6. Level of politicization <polit>

The level of politicization increased in the course of the negotiations, as the biosafety issue became more and more politically sensitive in Europe by the end of the 1990s. In the beginning of the negotiations, the issue was not characterized by a high level of politicization. A member state official described the first BSWG as follows: 'These negotiations were led by a sort of scientific ambiance, more than a political one'.

---

55 During the Luxembourg Presidency, the Netherlands and the UK even took over the Presidency's tasks in practice.

In the run-up to the Cartagena meeting, the negotiations – and the EU decision-making process – became more and more politically driven and politically sensitive in the EU member states. First, a growing public opinion resistance against GMOs – prompted by the first GMO on the European market, without labelling requirements – and an increasing NGO activity on the biosafety issue made the biotechnology debate in Europe extremely controversial (Bretherton, Vogler, 2000; Graff, 2002). Second, in various member states, green parties, which were more sensitive to biosafety, entered into government.[56] Third, various food safety crises across Europe gave rise to consumer concern for food safety. Fourth, simultaneous discussions about a *de facto* moratorium on GMOs[57] took place in the EU. Fifth, the precautionary principle and the increasing EU aim to get a strong precaution provision included in the Protocol, became politically sensitive, as the precautionary principle was considered the justification for the existing EC legislation in the field of biosafety, of which it became clear that the US would impose a WTO case against it (Falkner, 2007). In other words, the necessity of having included the precautionary principle in the Cartagena Protocol in order to justify its legislation, made this a very sensitive topic in the EU during the final negotiation session in Montreal.

7.7. Information asymmetry in favour of the agent <infag>

The EU negotiators, with the exception of the Commission as from the ExCOP, did not possess significantly more information about the negotiations than the principals did. All member state representatives describe the debriefings by Commission and Presidency in EU coordination meetings as 'effective' and 'working well'. If the EU negotiator had an information advantage vis-à-vis the principals, 'it was due to a selective debriefing because of practical reasons' (member state representative). None of the EU decision-makers stated that an EU negotiator had ever strategically concealed information. In most of the negotiation settings, member states could participate, as a result of which they usually had first-hand information. Only in the Group of Ten setting, used in Cartagena, principals were not able to attend the international negotiations.

7.8. Information asymmetry in favour of the principals <infprin>

The EU decision-making process was characterized by a large degree of transparency on the national preferences, and thus by a low degree of information asymmetry in favour of the principals. Member state and Commission representatives recognize that 'all cards were put on the table' and that 'everybody knew what the other had in mind'. A Commission official stated that the 'fallback

---

56   This was the case in France, Germany, Belgium and Italy.

57   In 1999, the EU decided not to approve any GMO product on its food or agricultural market: the so-called '*de facto* moratorium'.

positions of the main member states were known' and a Presidency representative claimed that 'as a negotiator, I had a pretty good idea of how far each member state could go. I knew what their instructions were'. That the same key players from each member state negotiated during a four-year period and that they were simultaneously negotiating the review of Directive 90/220/EEC in Brussels meant that the national positions were generally known. The EU compromise – and the way it was established between Commission and member states – in the final night of the Cartagena meeting is a good indication of the information symmetry with regard to the principals' bottom-lines. To save the negotiations, the EU proposed an ultimate compromise in which it gave in a lot on the precautionary principle to the Miami Group. Before the Commission tabled this compromise internationally, the Commission representative went back to the coordination meeting, on which he declared: 'I asked the member states to give me their bottom lines with respect to the requests of the Miami Group. All the member states gave me exactly their position'.

7.9. Degree of compellingness of the external environment <extcomp>

The degree of compellingness of the external environment was very high. Both in Cartagena and in Montreal, there was a strong will by the principals to have a Protocol adopted. Particularly in Montreal, one year after the collapse in Cartagena, the idea that a second collapse would mean the end of the Protocol played a large role in the EU. 'Coming home with an agreement was objective number one' (member state official). Environment Commissioner Wallström expressed the pressure on the EU and the member states to reach an agreement and to avoid jeopardizing the negotiation process as follows: 'We had come so far by this stage that the political cost of being the one who prevented the Biosafety Protocol from becoming reality would be enormous' (Wallström, 2002, p. 248). Moreover, in its conclusions of December 1999, the Environment Council 'recognise[d] the need for all participants to the negotiations to show the necessary flexibility in order to ensure a successful outcome in Montreal', 'emphasise[d] that every effort should be made to finalise the Protocol' and 'invite[d] the Commission and the Member States to continue to make every effort to bring the negotiations to a successful conclusion' (13854/99).

The main reason why the political cost of no agreement was extremely high in the member states was that a failure of the negotiations and the lack of an international instrument to deal with a highly controversial issue 'would return to be debated in the media or the streets' (Gupta, 2000, p. 27). Indeed, the protests at the WTO Ministerial Conference in Seattle (November/December 1999) were considered as an indicator of the sensitivity of the issue and generated concerns about the interplay between trade and environment (Falkner, 2000; Burgiel, 2002). Moreover, the member states feared that in case of a second collapse the GMO dossier would be taken over by the WTO, and that in this framework the environmental protection would get less attention.[58] An example of how the compellingness of the external environment influenced the principals' behaviour is the final agreement on the labelling issue, for which the EU had made a major concession to the Miami Group in order to get their approval of the compromise text. For the most maximalist member states (mainly Sweden, Austria and Denmark), 'the labelling provisions were really under our bottom line' (official from a maximalist member state). However, they agreed on the whole package to save the negotiations and the agreement.

7.10. Institutional density <instdens>

As mentioned above, the team spirit in the EU increased to a large extent during the course of the negotiations. The EU decision-making process took place in a very cooperative atmosphere. It is even regarded as a model of how the EU can maximize its international impact by operating as a team (Cameron, 2004). Hence, in particular in Cartagena, Montreal and the consultation processes in Vienna, the institutional density in the EU was extremely high. That was not so much the case during the first BSWGs.

Member state officials admit that the decision-makers definitely took into account the fact that they had to complete the whole process together (diffuse reciprocity): 'The Ad Hoc Group on Biosafety was made up with people who knew each other very well and who relied on each other during the negotiations.' Moreover, because of intense EU coordination processes, member states had a good insight into each other's positions and the rationale behind. A member state official expressed this mutual responsiveness norm as follows: 'We knew a lot about the other people and the positions in their capitals'. It is even acknowledged that some principals changed their positions to some extent because of discussions and interactions with other principals. For instance, some member states, which were not aware that labelling requirements could not be instantly realized, were persuaded by others on the practical difficulties of labelling GMOs.

Consensus and compromise striving always took a central place in the institutional environment of the EU coordination meetings. There was never a vote, or a threat to vote. Moreover, 'striving for a consensus was really the main objective of the EU coordinations' (member state representative). Finally, a large extent of trust between individuals, both among principals mutually and between principals and agents, grew as the decision-making process evolved. Representatives from the Commission and from negotiating Presidencies declared that they felt trusted by the member states, which was confirmed by the principals.

---

58   During the WTO Seattle Conference, members of the Miami Group had launched the idea to create a working group on biotechnology in the WTO, where the issues that were under discussion in the Cartagena negotiations could be dealt with (Bretherton, Vogler, 2000; Bail, Decaestecker, Jørgensen, 2002).

8. Stockholm Convention on POPs [STOCPOP]

8.1. The international negotiating process

The Stockholm Convention on Persistent Organic Pollutants (POPs) establishes a global regime for controlling chemical substances that are highly toxic to humans and animals, remain intact in the environment for many years, can travel long distances and accumulate in human and animal bodies. The Stockholm Convention aims to eliminate these POPs or to reduce their production, use, import, export, emission and disposal. Besides the elimination of 12 POPs – the so-called 'dirty dozen'[59] – the Convention foresees in a procedure to add new chemicals falling under its regulation. The Stockholm Convention was not the first international agreement to regulate these chemical substances. In 1998, a POPs Protocol to the regional (UNECE) LRTAP Convention[60] was negotiated. This UNECE POPs Protocol was signed and ratified by the EC and the 15 member states. Although the regional Protocol regulates more POPs, it is of a more limited scope than the (global) Stockholm Convention. However, it largely served as a blueprint for the Stockholm Convention.

From June 1998 to December 2000, five INCs under the auspices of UNEP were needed to reach an agreement on the Convention. Generally speaking, the negotiations were characterized by a cleavage between the Juscanz countries – led by the US – on the one side and the EU on the other side. Only on the financing issue were the EU and Juscanz united, as they wanted to use the existing facilities like the GEF, while G77 and China strived for a new financing mechanism (Vanden Bilcke, 2002). Although Canada can be considered as even more maximalist than the EU on a number of topics, the EU was taking a leading and demanding role in the negotiations (Delreux, 2009c). The 12 POPs of the Convention – and even more chemical substances – were already regulated in the EU. The EU started the negotiations with the position to ban 15 POPs. When it became clear that only the dirty dozen would be included in the Convention, it was more important for the EU to get progress in the negotiations, and the EU then strived for a flexible procedure to add new chemicals to the list afterwards. POPs management was already covered by EC legislation at the time of the negotiations. It was spread out over a prohibiting (79/117/EEC) and a biocide and pesticide directive (96/59/EC).

8.2. EU negotiation arrangement

Formally, the Presidency was EU negotiator during the POPs INCs: the Presidency spoke most of the time, made the official statements, etc. However, the EU

---

59   These 12 POPs are: PCBs, dioxins, furans, aldrin, dieldrin, DDT, endrin, chlordane, hexa chlorobenzene, mirex, toxaphene and heptachlor.

60   The LRTAP Convention is the Convention on Long-range Transboundary Air Pollution (1979).

negotiation arrangement evolved during the two and a half years of negotiations, as a result of a varying degree in the various Presidencies' strengths (Delreux, 2008; Delreux, 2009c).

In the first three INCs, the Presidency and the Commission – in INC 1 still on an informal basis – formed the EU negotiating team. Although POPs fell to a large extent under the competences of the EC – especially the trade components (export and import of POPs), but also the waste issues –, the Commission failed to get a mandate for INC 1. In the first INC, the rules of procedure for the international negotiations were established. As the Commission was not authorized to negotiate on behalf of the EC by that time, it was not included as a negotiator in the UNEP rules of procedure. However, the Presidency provided the Commission the opportunity to speak in plenary on issues covered by EC competences: the Presidency officially asked the floor and then passed it to the Commission. As a member state representative described it: 'Although the Presidency was the formal spokesperson for the EU, it did not mean that the Commission was a lame duck'. Hence, although the formal framework of the international negotiations did not allow the member states (Presidency) and the Commission to negotiate like the EU legal framework would suggest, in practice, they organized the division of labour according to the internal division of competences. By doing so, the Commission could play its role as EC negotiator. As member states did not make interventions in plenary during the first INCs, there was always a single European line, coordinated internally.

The EU negotiation arrangement changed at INC 4, where the Portuguese Presidency performed very weakly. As they could no longer rely on a well-performing agent, various member states began to take the floor separately, which stood out against the single voice strategy the EU had resolutely played in the previous INCs. The Commission then proposed a new negotiation arrangement based on a lead country approach. This basically meant that the different issues at stake – often the different articles of the draft Convention – were assigned to those member states that had most expertise and know-how on it and that wanted to take the lead in the international negotiations on behalf of the EU. In this new arrangement, individual member state experts took place next to the Presidency to negotiate on behalf of the EU on an issue they were specialized in. The Commission was sitting at the other side of the Presidency. When the EU took the floor in plenary, the Presidency immediately passed it either to the lead country expert or to a representative of the Commission. In such a situation, the experts *de facto* became a member of the Presidency team. This system was also used during INC 5 in Johannesburg because all players admitted that this arrangement had worked very well. Moreover, the EU's position was very precarious after INC 4. To adjust it, a strong negotiating system – being the lead country arrangement – was needed. In the preparation of INC 5, teams of two or three member states (sometimes together with the Commission) were formed.[61] In that stage, the division between national and Community competences did not play a role anymore. These teams prepared EU position papers, took the lead in the coordination meetings and negotiated internationally for these issues. Because of this evolution, the Commission's role also increased. In the last INC, the Commission got more space to act as agent than could be justified on the basis of the internal division of competences. 'Because the job of the Commission representatives was appreciated by the member states' (member state official), this was accepted by the principals. Hence, in plenary, the Presidency was the formal EU negotiator, but initially it gave the floor to the Commission for EC competence issues. When the negotiation process then came to an end, the Presidency did not only invoke the Commission, but also colleague member states, which acted as lead countries.

In the various contact groups, it was regarded useful that there would be more than one EU negotiator. Hence, member states were allowed to take the floor as well, but they had to respect the previously EU line. This differed from the negotiation arrangement used in plenary, where the member states did not take the floor. As contact groups by definition deal with specialized issues, every member state authorized the most experienced experts, irrespective from which member state they came, to negotiate on their behalf. In such settings, different European negotiators reinforced the common EU position by giving additional examples or arguments.

8.3. Discretion of the EU negotiator <disc>

All representatives – from the lead countries, the Presidencies and the Commission – recognize that 'the EU was very much a team. It was voluntary, and everybody somehow acknowledged we needed to cooperate, otherwise it would not have worked' (member state representative). The negotiation authority of the Presidency was never contested by the member states. Member state representatives admit that 'it was never challenged that the Presidency would be the EU negotiator' or 'it was never an option that member states would do it separately'. The fact that there was neither an authorization nor a mandate for the Commission before INC 1 may not be seen as the deployment of an *ex ante* control mechanism by the principals. The lack of mandate and authorization was caused by the Commission itself: it failed to initiate the proposal to the Council in time, probably due to practical problems (translation, etc.) in the Commission. The Council adopted a standard mandate for the Commission before INC 2. The only unique reference was to the UNECE POPs Protocol, with which the future Convention should be in line.

---

61   Examples of such teams were Portugal and the UK for the precautionary principle; Denmark, France and the UK for technical assistance and financial mechanisms; the Commission on dispute settlement and for the trade provisions; Belgium, Germany and the Commission for the waste issues; Sweden for liability; etc.

During the INCs, on the spot coordination meetings were organized several times a day. These meetings were sometimes attended by the candidate member states, if this was considered useful by the member states. This was not the case for the preceding preparatory meetings in Brussels, taking place both in a (formal) WPIEI context and in *ad hoc* POPs groups. Member states were able to attend the international negotiations in plenary and in the working groups. Only two or three times, a Friends of the Chair meeting was organized. As not every member state could be present in such a negotiation setting, it was mostly the Presidency, accompanied by the Commission and the appropriate lead country, representing the EU. A good illustration of the predominance of personal experience above the formal division of competences can be found in the role of the Commission in these Friends of the Chair meetings. Most of them dealt with financial issues, on which there is no Community competence, but it was mostly a well-experienced Commission representative leading the EU negotiation team on this matter.

In the negotiation stage, all EU positions were on paper, especially at the end, when the EU negotiator had comprehensive documents at his disposal, including multiple tables, with an EU position and an EU fallback position for every issue and for every article. Moreover, in the final stage of the negotiations, member states granted very little leeway to the EU negotiators with regard to some crucial points like the precautionary principle or the link with the Basel Convention.[62]

Neither during nor after the POPs negotiations did member states threaten not to ratify the Stockholm Convention. The ratification decision (2006/507/EC) – taken on the basis of article 175 TEC, and not on the basis of articles 95 and 175 TEC, like the Commission had proposed (COM(2003) 331) – did not generate a problem in the Council. However, there were some tensions among the member states on the implementation regulation that was simultaneously proposed by the Commission (e.g. on the internal EU procedure on how to propose new substances to be included on the Stockholm Convention). At the national level, 14 of the 15 member states (already) ratified the Convention.[63]

As the principals and the agents were really functioning as one EU team, there were no situations of the EU negotiators aiming to conquer additional autonomy vis-à-vis their principals. As a member state official stated: 'As we had such close contacts with each other, I cannot think of anything that the EU negotiator went somewhat beyond'. In INC 5, when a couple of Friends of the Chair meetings took

---

62   The EU position on precaution was not to get the terminology 'precautionary principle' in the Convention, but to get the procedure included that the lack of scientific evidence could not be considered as an obstacle to add a new POP to the Convention. Concerning the link to the Basel Convention (on the waste of POPs), the EU did not want to use the Basel mechanisms because of three reasons: the technical guidelines of the Basel Convention were not binding, the US were not a party to the Basel Convention, and the impression in the EU (in particular in the Commission) that the Basel committees were not working satisfactorily.

63   Only Italy did not ratify the Convention (dd. January 2011).

place, there was some concern among the member states, but this was not based on a fear of a shirking agent, but on 'the possibility that the spokesperson of the Presidency would not have enough power, experience or background' (member state representative). Hence, the various agents did neither aimed to conquer more autonomy nor did they abuse the autonomy granted by the principals.

8.4. Degree of preference homogeneity among principals <prefprin>

The preferences of the member states during the POPs negotiations were homogeneous. Certainly on all the technical issues (on chemicals), there were no significant differences among the member states' preferences, due to the fact that the 'dirty dozen' were already regulated in the EU (Delreux, 2009c). There were only minor differences between the member states, on issues like liability, financing or the institutional framework. Although member states very much looked in the same direction during the negotiations, member states like Denmark and Sweden had maximalist preferences on many issues (most likely because of the Nordic sensitivity for POPs). Also the Netherlands and Germany can be considered as having maximalist preferences, while the UK was somewhat more minimalist.

8.5. Degree of preference homogeneity between principals and agent <prefpa>

Five different Presidencies negotiated the Stockholm Convention on behalf of the EU during the five INCs: respectively the UK, Germany, Finland, Portugal and France. As these formal agents were also principals and as the degree of preference homogeneity among the principals was assessed as high, the distances between the principals and the agents can be considered small as well. Member state officials acknowledge that the various member states holding the Presidency – especially the UK, the French (rather on the minimalist side in the EU) and the Finnish and the German (on the maximalist side) Presidencies – 'ignored their national positions' during the INC in which they acted as agent. Portugal did not seem to have an identifiable position on the Stockholm Convention.

The Commission's preference may be situated on the moderate-maximalist side, quite similar to e.g. the German or the Dutch preference, yet not as strong as the preference of the Nordic member states. Analysing the degree of homogeneity between the member states on the one hand and the negotiating lead countries on the other hand was not really possible, as a lot of various (combinations of) member states took the lead on different issues.[64] However, as the member states' preferences were homogeneous, and as this homogeneity increased in the last

---

64   I did not have appropriate data on the preference distributions in the EU on the approximately 15 issues on which a lead country approach was established. To analyse the degree of preference homogeneity between principals and lead countries, both data on the member states' preferences on that particular issue and data on the lead country's preferences are needed.

stages of the negotiations (when the lead country approach was used), the degree of preference homogeneity between member states and lead countries was very high. As a consequence, all agents' preferences were situated within the – rather narrow – range of principals' preferences.

8.6. Level of politicization <polit>

The level of politicization of the POPs negotiations was moderate in the EU member states. The negotiations were attended by NGOs and the media paid considerable attention. Mainly the precautionary principle and the aim of the EU to get this principle included in the Convention, made it relatively politicized. No ministers from the EU side, attended the INCs. The 12 substances under discussion were already regulated in the EU and European chemicals policy already went further.

8.7. Information asymmetry in favour of the agent <infag>

Member state officials, coming from both countries holding and countries not holding the Presidency, affirm that the 'flow of information was good', 'the debriefing effective', and 'everyone could know everything, if he or she just asked it'. This last statement confirms that, like in most negotiations, the information a member state has at its disposal mainly depends on the involvement and the interest of that member state in the international negotiations. The relation between these closely involved principals and the agent was 'very transparent, because there was no hidden information' (member state representative).

8.8. Information asymmetry in favour of the principals <infprin>

Just like the agent had no information benefit vis-à-vis the principals, the principals did not have non-transparent fallback positions vis-à-vis the negotiating Presidencies: 'Nobody came with surprises. There were no hidden agendas or late indications of certain positions. It was in a quite good spirit and openness' (official from a member state holding the Presidency at an INC). This was confirmed by other member state and Commission representatives, although one national official noted that 'on the broad issues, we were all open, but when you come to the detailed issues, there can be fallback positions'.

8.9. Degree of compellingness of the external environment <extcomp>

A lot of member states were closely involved in the negotiations. Not only did five of the fifteen member states hold the EU Presidency – and, as a consequence, were the EU negotiator –, but every member state could attend the plenary and the various contact group meetings. Hence, they were in an appropriate position to experience the pressure of the international negotiations. As the EU had a demanding position at the international level, member states wanted to avoid blowing up the negotiation process, in particular at INC 5. The principals experienced, as noted by a member state official, a pressure in the sense of 'we have to reach an agreement, we have started a process and we cannot stop it now'.

8.10. Institutional density <instdens>

Negotiators from the member states, the Presidencies and the Commission acknowledge that the relation between the EU negotiators and the member states was not conflictual, but 'very cooperative'. A Commission official stated it like this: 'It was really joined, we were in the same boat. There was, certainly in Johannesburg [INC 5], a feeling among the Europeans of "if you touch him, you also touch me".' In particular at the last INC, there was a strong feeling of solidarity in the EU, which was even described as 'one European family'. This was a reaction on the debacle with the Portuguese Presidency in INC 4, as a result of which the position of the EU was remarkably weakened, mainly vis-à-vis Juscanz. The desire to 'fight back to the Americans' (Commission representative) certainly triggered the cooperative nature of the EU decision-making process towards the end of the negotiations.

The institutional environment in the EU was quite dense during the negotiations on the Stockholm Convention. Member state officials noted that 'there was a good understanding', and that 'the power of the argument played a role' (mutual responsiveness). Moreover, diffuse reciprocity seemed to matter, as a member state representative declared: 'Member states respected the fact that they had to be able to cooperate afterwards'. The striving for a consensus in the EU was large as well. In the final hours of the negotiations at INC 5, an EU coordination meeting with only one representative per member state was organized to solve a package of issues that were very important for the EU (the precautionary principle, finances and the link with the Basel Convention).[65] Voting did not occur in this limited EU coordination meeting, but 'a vote was very very close' (member state representative). The fact that in such a meeting, in the final hours before the end of an international negotiation, member states opted not to have a formal vote, but to solve the issues by looking for a consensus shows the high degree of institutional density in the EU. The final indicator of institutional density was the high degree of trust of the principals in their agents. However, not all agents were trusted to the same degree. The Commission and most Presidencies could rely on the member states' confidence, although the less well-performing Presidencies (mainly the Portuguese) were not entrusted. The lead countries were trusted as well, but this is more evident as they were chosen because of their active involvement and their excellent experiences on the issues.

---

65   The importance of these three issues is shown by the fact that they ware the main substantive points of the Council Conclusions on POPs of 7 November 2000, preceding INC 5 (12924/00).

9. SEA Protocol [SEA]

9.1. The international negotiating process

The Protocol on Strategic Environmental Assessment is concluded under the 1991 UNECE Espoo Convention, which deals with Environmental Impact Assessment (EIA) in a transboundary context. The Espoo Convention stipulates that its parties have to inform each other when they are developing projects, which could have a negative environmental transboundary impact. While the Espoo Convention only applies to projects, the SEA Protocol extends its scope to plans and programmes. The main points of discussion during the negotiations were whether the scope of the Protocol would even be extended to policies and legislation, whether access to justice and broader public participation provisions had to be included in the Protocol, and to what extent health impact assessment had to be part of the environmental assessment.

The SEA Protocol was negotiated under the auspices of UNECE between May 2001 and January 2003. In the second MOP of the Espoo Convention (February 2001), an Ad Hoc Working Group was established to negotiate the Protocol. After eight sessions, the Working Group finalized its work, and the Protocol was signed at the fifth ministerial 'Environment for Europe' conference in Kiev in May 2003. MOP 2 of the Espoo Convention assessed that there was a political momentum to extend the Espoo Convention to include plans and programmes, because this had also been done at the EU level, where the EIA legislation existing since (85/337/EEC; amended by 97/11/EC; and by 2003/35/EC) was just transformed to an assessment of plans and programmes in the so-called SEA Directive (2001/42/EC). As a result, UNECE's aim was to pick up on this process (from EIA legislation that covers projects to SEA legislation covering plans and programmes) at the international (regional) level. The negotiations at the EU level on the SEA Directive had taken more than four years. They were characterized as tough and difficult negotiations between member states, the Commission and the European Parliament (Feldmann, 2004). From an EU perspective, the moment to start similar negotiations at the international level right after the finalization of the negotiations of the SEA Directive was unfortunately chosen. A member state representative expressed this as follows: 'It was extremely awkward by UNECE. It made real negotiations quasi-impossible.' Indeed, the fact that the EU member states were asked to start again the negotiations they had just finalized at the EU level on the one hand, and the fact that they had just started to implement the SEA Directive on the other hand, meant that the EU had very limited flexibility to negotiate (Delreux, 2009b). From the EU perspective it was clear that the scope of the SEA Directive could not be opened through UNECE negotiations (Feldmann, 2004). As the other powerful UNECE members Canada, the US and Russia did not play a (significant) role in these negotiations[66] and as the EU was thus the largest negotiation block, this EU position impeded real and open negotiations.

Norway, the Espoo Secretariat and some Central and Eastern European countries – mainly Poland and the Czech Republic – had the most maximalist positions at the international level, as they strived for a comprehensive Protocol, which would even go further than the Aarhus Convention in the areas of public participation and access to justice. NGOs, which were able to participate in the negotiations, took a similar position. A majority of the EU member states (mainly with the exception of Italy, see further) had the most minimalist positions at the international level, as they did not want to go beyond the SEA Directive. The final text of the SEA Protocol does not diverge explicitly from the SEA Directive. In that sense, the EU has won its points in these negotiations. However, there are some provisions in the Protocol going beyond the Directive, but these are formulated in soft law language, so that they are considered not to require a review of the SEA Directive.

9.2. EU negotiation arrangement

The EU negotiation arrangement during the SEA Protocol negotiations was rather unstructured. During the first half of the negotiations, i.e. until approximately the fifth negotiation session, member states spoke freely in their capacity of UNECE member states. From the second session onwards, the Commission was authorized and had a mandate to act as a negotiator for issues covered by the SEA Directive. The other issues, such as the application of SEA on policies and legislation, or the extension of the field of application to health were still covered by national competences. However, in practice, it was not only the Commission taking the floor for the issues covered by the Directive: member states also spoke when these issues were touched upon. In such cases, the member states usually supported the position expressed by the Commission, but they sometimes also expressed preferences that diverged from the position the Commission had presented. Hence, during the first half of the negotiations, both Commission and member states negotiated at the international level, but the formal division of labour on the basis of competences was not followed in practice, which contributed to an unstructured negotiation arrangement (Delreux, 2009b). Moreover, the Commission also took the floor to speak in its capacity of the Commission, and not only on behalf of the member states.

From the fifth negotiation session onwards, the EU negotiation arrangement became even less structured: the principals started to strive for a common position, for which they delegated negotiation authority to the Presidency. As a consequence, the member state part of the EU negotiation arrangement, which was expressed

---

66 The US never participated in the negotiations, Canada was only occasionally present, and Russia negotiated in the first sessions, but left the negotiations at an early stage.

by the member states separately in the sessions before, became more and more coordinated and now expressed by the Presidency at the international level.[67] However, although the principals' positions became more and more coordinated, there was no single EU voice as the member states still took the floor as well. This led to a situation in which the Commission (for issues falling under the SEA Directive), the Presidency (for issues not directly linked to the SEA Directive and on which the member states had a common position) and the member states themselves (for issues on which there was no common position, but also for the issues expressed by the Commission or by the Presidency) took the floor at the international level.

Because several actors played a negotiating role and because these actors did not strictly respect their roles, the EU negotiation arrangement was rather chaotic. First, the Commission was the formal agent, on the basis of a mandate, for the issues falling under EC competence. However, the Commission also took the floor in its own capacity as UNECE observer when other issues were discussed. The Commission thus acted as agent – although not exclusively – for the EC part of the negotiations, and as the Commission – or as 'a kind of sixteenth member state' (member state official) – for the other part. Second, the Presidency sometimes acted as agent for issues on which the member states had a coordinated position, but it could also take the floor as a member state on other issues. And finally, the member states intervened to support or even sometimes to oppose the Commission or the Presidency or to speak freely for issues on which there was no EC competence or no common member state position. A Commission representative summarized it as follows: 'Who wanted to speak, spoke'. However, this happened less frequently when the negotiations came to an end.

9.3. Discretion of the EU negotiator <disc>

The member states seem to have been rather reluctant to delegate negotiation authority to the Commission. In the beginning, the relation between some of the member state representatives and the Commission officials was tense and difficult. As indicated, this can be explained by the internal decision-making process on the SEA Directive, which took place just before the UNECE negotiations started. In this stage, the Commission aimed to convince the member states that 'we are not your enemy anymore, we are your voice' (Commission official). Before the second negotiation session, the principals granted a mandate to the Commission. The mandate stipulated that the outcome of the negotiations had to be consistent with the SEA Directive and with the relevant parts of the Aarhus Convention, to which the EC was going to become a party. It was clear that the mandate did not grant any margin to go beyond the SEA Directive. Also, when the negotiations evolved, the

---

67 It also happened that the 'coordinated position', as expressed by the Presidency, was not agreed upon by all member states, but only by a majority of them. The Presidency thus also expressed positions 'on behalf of the majority of the member states'.

member states did not grant additional autonomy to the Commission. The Directive was – and remained – the red line that the agent was not allowed to cross. By the end of the process, the Presidency was granted negotiation authority – often by an *ad hoc* majority of the member states – because this allowed the member states that shared a common position to express their views more strongly.

During the negotiations, the principals were free to attend every negotiation setting at the international level. There were no meetings with restricted participation. During the weeks of UNECE negotiations, coordination meetings were organized on a daily basis. Moreover, the member states asked the Commission twice to arrange a coordination meeting in Brussels between two negotiation sessions. Moreover, in-between the negotiation sessions, an informal group of the most involved member states (being also the member states with the most minimalist preferences) had regular contacts by e-mail.[68] From the second half of the negotiation process, the then candidate member states were allowed to attend the EU coordination meetings. These countries were not yet formally bound by the SEA Directive, but they were in the stage of implementing the *acquis communautaire*, including this Directive. Initially, they had more maximalist positions (in particular Poland and the Czech Republic), but they evolved more and more in the minimalist direction, as they were involved in the EU coordination meetings. From that point on, the EU took a strong position at the international level, as the number and relative bargaining power of the external negotiation partners had become very limited. In the negotiation stage, some member states threatened to leave the negotiations. However, this may not be seen as an indication of an involuntary defection threat vis-à-vis the Commission. The principals' dissatisfaction had nothing to do with the behaviour of their agent, but rather with the role of the international chairman, the Espoo Secretariat and the way the international negotiation process was managed.

The SEA Protocol was ratified by the EC in 2008 (2008/871/EC) and national ratifications in the member states are currently going on.[69] In the build-up to the EC ratification, the main question at the Commission, which has to propose such a decision to the Council, was whether the SEA Protocol required an amendment of the SEA Directive or not. As mentioned, some provisions in the Protocol that go beyond the Directive were formulated in soft law wording. This was accepted by the EU in the negotiation stage, but before proposing the ratification decision, the Commission seemed to be somewhat more cautious as it wanted to prevent a situation where ratification by the EC would open up discussions on whether these soft law provisions required an amendment of the SEA Directive.

---

68   This group mostly consisted of the Commission, Austria, Germany, Finland, Sweden, France and the UK.

69   As of January 2011, 14 member states have ratified the SEA Protocol: Austria, Bulgaria, Czech Republic, Estonia, Finland, Hungary, Germany, Luxembourg, the Netherlands, Romania, Slovakia, Slovenia, Spain and Sweden.

Both member state and Commission officials acknowledge that none of the agents went beyond his room of manoeuvre. Consequently, the agents did not conquer more autonomy than the autonomy granted by the principals.

9.4. Degree of preference homogeneity among principals <prefprin>

Member states' preferences were rather heterogeneous in the course of the SEA Protocol negotiations. Although most principals had similar (minimalist) preferences, some of them wanted to go further and strived for a more demanding Protocol. The preference of the majority of the member states was that the provisions of the Protocol should be as close as possible to those of the SEA Directive. However, there was no agreement on this in the EU as the maximalist member states like Italy, Denmark, Finland, Belgium (to a lesser extent) and the Netherlands (in particular in the beginning of the process) strived for a broader field of application (i.e. including health in the field of application), for stronger public participation provisions, and some of them for the inclusion of an access to justice article. Especially Italy is considered to have had the most maximalist preferences, sometimes even leading to an isolated position in the EU. By contrast, France, Germany, the UK and Sweden were on the minimalist side of the preference spectrum. They aimed to avoid any deviation from the Directive in the Protocol. In the second half of the negotiations, the degree of preference heterogeneity decreased in the EU, as some of the maximalist member states moderated their preferences: a change of government in the Netherlands, Denmark and Italy led to a more homogeneous preference distribution among the member states, as they all went into a more minimalist direction ('the Directive and only the Directive').

9.5. Degree of preference homogeneity between principals and agent <prefpa>

The Commission's preference on the SEA Protocol was similar to the preference of the most minimalist member states: it aimed to avoid that the SEA Directive's field of application had to be expanded. The role of the Commission was to make sure that the negotiated Protocol would not go beyond the EC legislation. As a consequence, the degree of preference homogeneity between principals and Commission was high at the end of the process. In the beginning, however, the agent's preferences only corresponded to those of the minimalist member states, as member states like Italy did not consider the Commission's preferences to be in line with their own preferences.

During the final negotiation sessions, the Presidency also acted as agent. At that time, Denmark (which was the Presidency during the sixth and the seventh session) already had a moderate preference after its change in government, and Greece (as the Presidency during the short eighth session) did not really have an outspoken preference on the Protocol. Consequently, to the extent the Presidencies acted as agent, their preferences did not diverge from the principals'.

9.6. Level of politicization <polit>

The negotiations on the SEA Protocol were not characterized by a high level of politicization. The negotiations were conducted by experts, some of them often operating without national instruction. Moreover, there was no media attention for the negotiations. With the exception of the UK (on the possible extension to the level of policies), Germany and Austria (on the implications the Protocol would have on their federal state structure), 'the Protocol was not considered important in the member states' (member state official).

9.7. Information asymmetry in favour of the agent <infag>

The Commission – and the Presidency, in the final stage – did not enjoy an information benefit vis-à-vis the member states. Member state and Commission representatives witnessed that 'the EU negotiator never knew more than the member states'. Moreover, there were no situations in the negotiations that created an information asymmetry in favour of the agent (e.g. negotiation settings in which only the EU negotiator was present).

9.8. Information asymmetry in favour of the principals <infprin>

During the SEA Protocol negotiations, the EU negotiator was well acquainted with the range of possible agreements that the member states could ultimately accept. In the international negotiations, the agent and the principals were mostly represented by the same officials who had negotiated the SEA Directive, which led to a situation where 'everyone knew everyone's positions' (member state representative). Another member state official described this situation as follows: 'There were no hidden cards. The Commission knew the position of every member state.'

9.9. Degree of compellingness of the external environment <extcomp>

The external environment was not very compelling, although the consideration that 'we had to have something to propose to the ministers' (member state official) at the ministerial conference in Kiev, where the SEA Protocol was planned to be signed, created some time pressure. However, the general degree of compellingness was low, in particular since the candidate member states had joined the EU coordination meetings. From that moment on, the external negotiation partners were only 12 (mostly CIS) countries, which meant that the EU *de facto* dominated the negotiations without experiencing a large pressure from its international partners to come to an agreement. A member state official stated that 'in practice, it was only Norway, Switzerland, the Secretariat and the NGOs'. The low degree of external compellingness is also illustrated by the fact that the member states and the Commission considered to leave the negotiation at the fifth session because of its mismanagement (see above).

9.10. Institutional density <instdens>

The EU decision-making process with regard to the SEA Protocol negotiations was not characterized by a dense institutional environment. According to member state and Commission officials, 'the EU was not really a team' or 'there was no big EU feeling'. Although the officials who negotiated the SEA Protocol frequently met each other, not only during the eight negotiation sessions, but also in the Council Working Groups during the years before and in the EIA-SEA expert committee, none of them assessed the diffuse reciprocity norm as important. Nor does mutual responsiveness seem to have played an important role in the EU decision-making process, as officials acknowledge that justifications of national preferences – even in the small informal group of member states that closely stayed in touch in-between the negotiation sessions – hardly ever occurred and that it did not influence other member states' preferences. It was particularly difficult to understand the Italian position, as 'they did not give us the information on how they would implement a broad Protocol' (member state official). There was never a strong striving for consensus and compromise in the EU coordination meetings. At the end of the negotiation stage, this striving somewhat increased, but certainly for negotiation session 1 to 5, officials did not discover a strong willingness to reach a common position. Finally, member state and Commission representatives state that there was not a large degree of trust from the principals in their agent. The experiences of the decision-making process on the SEA Directive, which were characterized by a member states versus Commission situation, still had an impact, certainly in the beginning of the negotiations. A member state official even expressed it like this: 'There was a lack of trust. The member states did not trust the Commission, and vice-versa.'

10. Comparative overview

Table 4.2 summarizes the empirical findings of the cases. As the cases are presented in the columns of the table, each column can be read as an 'identity kit' of the case. By representing the variables and their dimensions in the rows, reading row by row makes it possible to compare the eight cases on their different aspects.

The EU decision-making process with regard to the negotiations leading to the Kyoto Protocol, the Aarhus Convention, the Cartagena Protocol, the Stockholm Convention and the SEA Protocol are characterized by an evolution in time. I present this evolution by distinguishing between two stages in the decision-making process. In Table 4.2, these two stages are indicated by a '[1]' for the first and by a '[2]' for the second stage. For each of these five cases, I defined a turning point in the process:

- the switch from the Dutch to the Luxembourg Presidency in June 1996 for the *Kyoto Protocol* negotiations;. During the first period, the strong Dutch Presidency acted as EU negotiator, the international negotiations took place in Subsidiary Bodies that could be attended by the member states, and the principals' preferences were more heterogeneous. The second period was characterized by a weaker Luxembourg Presidency, as a consequence of which the troika became the agent, by a final negotiation session in Kyoto where the troika often represented the EU without the member states being able to attend the international negotiations, and by increasingly homogeneous preferences among the principals (although still rather heterogeneous);
- the beginning of the eighth negotiation session of the *Aarhus Convention* negotiations in December 1997, because, from then on, the Commission was authorized to represent the member states;
- the change of negotiation arrangement in the *Cartagena Protocol* negotiations at the end of the sixth Biosafety Working Group in February 1999. From that moment on, the Commission was *de facto* the only agent, the team spirit in the EU had increased to a very high degree, limited negotiation settings were held at the international level, member states' preferences became more homogeneous and the degree of politicization had increased enormously in the EU;
- the switch from the Finnish to the Portuguese Presidency before INC 4 of the *Stockholm Convention* negotiations (March 2000), which meant a switch from a period with strong Presidencies to a period with weaker Presidencies. This caused a change in the EU negotiation arrangement, as in the second period of the Stockholm Convention negotiations, the lead country approach was used. Moreover, the institutional environment in the EU was denser in the second period than in the first;
- the fifth negotiation session of the *SEA Protocol* negotiations in May 2002. At this point, the EU negotiation arrangement became somewhat more structured, the Presidency took an agent role and the institutional norms began to play a more significant more than before.

Table 4.2 Comparative overview of the cases

1. Characteristics MEA and international negotiations

	CCD	AEWA	KYOTO	ARHUS	PIC	CART	STOCPOP	SEA
			([1]: 02/1997-06/1997; [2]: 07/1997-12/1997)	([1]: session 1-7; [2]: session 8-10)		([1]: BSWG 1-6; [2]: ExCOP-ExCOP-bis)	([1]: INC 1-3; [2]: INC 4-5)	([1]: session 1-4; [2]: session 5-8)
full name	United Nations Convention to Combat Desertification	Agreement on the Conservation of African-Eurasian Migratory Waterbirds	Kyoto Protocol to the United Nations Framework Convention on Climate Change	Convention on Access to Information, Public Participation in Decision Making and Access to Justice in Environmental Matters	Rotterdam Convention on the Prior Informed Consent Procedure for Certain Hazardous Chemicals and Pesticides in International Trade	Cartagena Protocol on Biosafety to the Convention on Biological Diversity	Stockholm Convention on Persistent Organic Pollutants	Protocol on Strategic Environmental Assessment to the Espoo Convention
purpose	combating desertification and soil degradation in a number of regions	conserving migrating waterbirds in Africa and Eurasia	combating climate change by imposing quantified emission targets and flexible mechanisms	granting rights to citizens in environmental decision-making	taking informed decisions on import of chemicals	regulating transboundary movement of certain GMOs	eliminating or reducing POPs	imposing strategic environmental assessment for projects, plans and programmes
period negotiations	05/1993-06/1994	06/1994-06/1995	02/1997-12/1997	06/1996-03/1998	03/1996-03/1998	07/1997-01/2000	06/1998-12/2000	05/2001-01/2003
auspices	UN	UN	UN	UNECE	UN + FAO	UN	UN	UNECE
overview positions international level	SQ — Juncura — EU — G77 ⎫ =WEOG	SQ — African + former CIS countries — EU	SQ — Australia — US — Japan — EU — non-Annex II countries* ⎪ *had no commitments to make	SQ — Russia — CEE — EU — Norway + Poland — NGOs	SQ — Juncura — EU	SQ — Miami Group — EU — CEE Group — Compromise Group — Like Minded Group	SQ — Juncura — EU — (Canada) — Norway — Switzerland	SQ — EU* — Norway — Czech Rep. — NGOs *with some exceptions
position EU	• no new financing • no use of the GEF for desertification	level of conservation as high as possible, within the boundaries of the Bird and Habitat Directives	• introducing the bubble concept • binding emission reduction target of 15% • common coordinated policies and measures	no clear common EU position	• making the PIC procedure internationally legally binding • framework convention	• broad scope • AIA, labelling and documentation included • precautionary principle • opposed to savings clause	• regulating (initially) 15 and (later on) 12 POPs • flexible mechanism to add new POPs	not going beyond the SEA Directive

	CCD	AEWA	KYOTO ([1]: 02/1997-06/1997; [2]: 07/1997-12/1997)	ARHUS ([1]: session 1-7; [2]: session 8-10)	PIC	CART ([1]: BSWG 1-6; [2]: ExCOP-ExCOP-bis)	STOCPOP ([1]: INC 1-3; [2]: INC 4-5)	SEA ([1]: session 1-4; [2]: session 5-8)
2. EU negotiation arrangement and EU representation								
agent	Presidency	Commission	[1] Presidency [2] troika	[1] no agent [2] Commission	Commission	[1] Commission + Presidency [2] mainly Commission, Presidency on limited number of issues	[1] Presidency + Commission [2] lead countries + Commission	[1] Commission [2] Commission + Presidency
principals taking the floor?	yes	yes	[1] occasionally [2] no	yes	no* * except for some non-substantial issues	no	no	yes
single voice?	mostly	yes	yes	no	yes	yes	yes	no
single mouth?	no	no	[1] yes [2] no	no	yes* * except for some meetings of the contact groups	no	no	no
existing EC legislation	—	79/409/EEC 92/43/EEC	—	85/337/EEC 90/313/EEC 96/61/EEC	2455/92/EC	90/219/EEC 90/220/EEC	79/117/EEC 96/59/EC	2001/42/EC
3. Discretion <disc>								
authority contested?	no	no	no	yes	no	[1] yes [2] no	no	yes
mandate?	no	yes	no	[1] no [2] yes	yes	yes	yes* * as from INC 2 onwards	yes
clear instructions?	yes	yes	yes	no	yes	yes	yes	yes
coordination meetings?	yes* * : frequent WEOG coordinations	yes	yes	[1] no/very little [2] yes	yes	yes	yes	yes

	CCD	AEWA	KYOTO ([1]: 02/1997-06/1997; [2]: 07/1997-12/1997)	ARHUS ([1]: session 1-7; [2]: session 8-10)	PIC	CART ([1]: BSWG 1-6; [2]: ExCOP-ExCOP-bis)	STOCPOP ([1]: INC 1-3; [2]: INC 4-5)	SEA ([1]: session 1-4; [2]: session 5-8)
3. Discretion <disc> (cont.)								
member states attending?	yes	yes	[1] yes [2] no* *i.e. the meetings with US and Japan in Kyoto	yes	yes	[1] yes [2] initially: no; later on: yes	yes	yes
limited negotiation settings?	yes	no	[1] no [2] yes	yes	yes	[1] no [2] yes	yes	no
threat of involuntary defection?	no	no* *except for some federal member states at the end of the last session	no	no* *except for Germany and its domestic ratification	no* *except for one national representative at the end	no	no	no
EC ratification decision	98/216/EC	2006/871/EC	2002/358/EC	2005/370/EC	2003/106/EC, replaced by 2006/730/EC	2002/628/EC	2006/507/EC	2008/871/EC
agent conquering more autonomy?	no	no	[1] no [2] yes	[1] *not applicable* [2] no	no	no	no	no
4. Degree of preference homogeneity among principals <prefprin>								
homogeneous/ heterogeneous	homogeneous	heterogeneous	heterogeneous	heterogeneous	homogeneous	[1] heterogeneous [2] homogeneous	homogeneous	heterogeneous
overview positions EU level	SQ —DE/UK —FR/NL —(ES)* * on Annex 4 until INCD 3	SQ —FR —UK —DE/NL	SQ —FI/ES/IT/PT —NL/UK —AT/DE/DK/SE	SQ —DE —FR/UK —NL —BE/DK/IT	SQ —UK —DK/FI/NL/SE	[1] SQ —DE/FR —NL/UK —AT/DK/SE [2] SQ —NL/UK —AT/DE/DK/ FR/SE	SQ —UK —DE/NL —DK/SE	SQ —AT/FR/DE/ SE/UK —BE —DK/FI —IT

	CCD	AEWA	KYOTO ([1]: 02/1997-06/1997; [2]: 07/1997-12/1997)	ARHUS ([1]: session 1-7; [2]: session 8-10)	PIC	CART ([1]: BSWG 1-6; [2]: ExCOP-ExCOP-bis)	STOCPOP ([1]: INC 1-3; [2]: INC 4-5)	SEA ([1]: session 1-4; [2]: session 5-8)
5. Degree of preference homogeneity between principals and agent <prefpa>								
homogeneous/ heterogeneous	homogeneous	homogeneous* * except for France	[1] heterogeneous [2] heterogeneous, but to a smaller degree than in [1]	[1] not applicable [2] homogeneous	homogeneous	homogeneous	homogeneous	homogeneous
6. Degree of politicization <polit>								
degree of political sensitivity in EU	very low	very low* * except for France	very high	low	very low	[1] low [2] very high	[1] moderate [2] high	low
7. Information asymmetry in favour of the agent <infag>								
information benefit for agent?	no	no	[1] no [2] yes	no	no	yes	no	no
negotiation settings with only agent and no principals?	no	no	[1] no [2] yes	no	no	[1] no [2] initially: yes*; later on: no *: 'Group of Ten'	yes	no
8. Information asymmetry in favour of the principals <infprin>								
information benefit for principals?	no	no* * except for France	yes	no	no	no	no	no
9. Degree of compellingness of the external environment <extcomp>								
external compellingness	low	very low	very high	low	moderate	[1] high [2] very high	high	low
scope of the negotiations	global	intercontinental: European, African and CIS countries	global	regional: European and CIS countries	global	global	global	regional: European and CIS countries

	CCD	AEWA	KYOTO [1]: 02/1997-06/1997; [2]: 07/1997-12/1997)	ARHUS [1]: session 1-7; [2]: session 8-10)	PIC	CART [1]: BSWG 1-6; [2]: ExCOP-ExCOP-bis)	STOCPOP [1]: INC 1-3; [2]: INC 4-5)	SEA [1]: session 1-4; [2]: session 5-8)
10. Agent as subset of the principals <agprin>								
member state(s) acting as agent?	yes (Presidency)	no	[1] yes (Presidency) [2] yes (troika)	no	no	partly (Presidency, besides the Commission)	[1] partly (Presidency, besides the Commission) [2] partly (lead countries, besides the Commission)	[1] no [2] partly (Presidency, besides the Commission)
11. Institutional density <instdens>								
general degree of institutional density	moderate	moderate	[1] high [2] moderate	very low	high	[1] low [2] very high	[1] high [2] very high	[1] very low [2] moderate
diffuse reciprocity	high	high	very high	very low	high	[1] low [2] high	high	very low
mutual responsiveness	very low* * considered as unnecessary in this process	very high	very high	very low	high	[1] low [2] high	high	very low
compromise and consensus striving	low	high	high	very low	very high	[1] moderate [2] very high	[1] high [2] very high	[1] very low [2] moderate
trust	high	low	[1] high [2] moderate	[1] *not applicable* [2] high	very high	[1] low [2] very high	high	[1] very low [2] moderate
avoiding process failure	high	moderate	very high	low	high	[1] moderate [2] very high	high	low

11. Conclusion

The eight case studies presented in this chapter clearly show that the EU negotiator's discretion largely varied in the various decision-making processes studied. While there was e.g. no discretion for an agent in the first stages of the Aarhus Convention negotiations, the agent's discretion was high in the final stages of the Kyoto Protocol negotiations. The discretion enjoyed by the EU negotiator in the other cases can be positioned in between those two.

The next chapter (Chapter 5) explains this variation and identifies the (combinations of) conditions leading to a particular degree of discretion. It therefore first distances itself somewhat from the in-depth empirical data presented in the current chapter, since the data will first be simplified in order to analyse it systematically. This simplification implies a dichotomization of the data, leading to a grouping of the cases on each of the different variables. On the basis of a systematic and formalized comparison of this dichotomized data, explanations for the EU negotiator's discretion are identified. Then, the case studies presented in the current chapter are reintroduced in order to describe the causal mechanisms behind these explanations.

Also Chapter 6 relies on the eight case studies presented above. Unlike Chapter 5, which is centred around the explanation of the outcome variable 'discretion', Chapter 6 examines the condition variables and draws more general conclusions on the EU decision-making process as such. These findings are based on the observation that, though their considerable degree of variation, the eight EU decision-making processes presented in the current chapter share common characteristics on how their main determinants, the condition variables, function in practice and on how these decision-making processes are conducted on the field.

Chapter 5
Explaining the EU Negotiator's Discretion

1. Introduction

In the previous chapter, I described the cases in terms of the variables and I systematically compared them in Table 4.2. In the current chapter, I convert this data to values suitable for being analysed via QCA. This requires the positioning of a threshold between values with a low degree and values with a high degree of a certain variable. In Section 2, the main features of the method for data analysis, Qualitative Comparative Analysis (QCA), are presented. In the third section, the empirical data presented in the previous chapter is dichotomized in order to prepare it for a QCA procedure. Section 4 then analyses this QCA-ready data by means of Boolean minimization. In this way, it can be explained which combinations of conditions explain a certain degree of discretion. In the fifth section, these explanations (in the form of minimal formulae) are interpreted in the light of the empirical data. This way, in order to understand the results of the QCA procedure, the data described in the previous chapter is brought back into the analysis.

2. Systematically analysing an intermediate-N situation with QCA

2.1. From eight empirical cases to 18 technical cases

In Section 4.2 of Chapter 3, I presented the eight cases studied in this book and I justified their selection. These are eight 'empirical cases': EU decision-making processes with regard to international negotiations leading to an MEA, which are defined in time and in substance. However, in the data analysis stage in this chapter, a case is seen as a unit of analysis: as a 'technical case'. Here, a case is not defined substantively, but only technically: a particular configuration of values on condition variables and a value on the outcome variable. Some of the selected 'empirical cases' are split up into multiple 'technical cases' because of a twofold reason.

On the one hand, I observed considerable variation in time on the EU negotiation arrangement or at least one of the variables. Those decision-making processes – which entail two different configurations of variables, caused by the evolution in time in the decision-making process – were split in technical cases: suffix '_1' refers to the first stage of the process, suffix '_2' to the second stage. By decoupling these empirical cases, I introduce a time dimension in the cases (Clément, 2004). The time periods of these cases are presented in the previous chapter, more in particular in its tenth section and in Table 4.2.

On the other hand, in some of the decision-making processes, both the Commission and the Presidency or a lead country acted as agent. This also led to a technical splitting-up of some cases: suffix '*a*' is used for a technical case where the discretion of the Commission (not a subset of the principals) is explained, while suffix '*b*' is used for the cases that are used to explain the discretion of the Presidency or a lead country (which are a subset of the principals). Consequently, the analysis is performed with 18 technical cases, presented in Table 5.1.

Table 5.1   Overview of the technical cases used in the QCA procedure

Case code	Technical case: EU decision-making process with regard to the...
CCD	UNCCD negotiations
AEWA	AEWA negotiations
KYOTO_1	1st stage of the Kyoto Protocol negotiations
KYOTO_2	2nd stage of the Kyoto Protocol negotiations
ARHUS_1	1st stage of the Aarhus Convention negotiations
ARHUS_2	2nd stage of the Aarhus Convention negotiations
PIC	Rotterdam PIC Convention negotiations
CART_1a	1st stage of the Cartagena Protocol negotiations, negotiated by the Commission
CART_1b	1st stage of the Cartagena Protocol negotiations, negotiated by the Presidency
CART_2a	2nd stage of the Cartagena Protocol negotiations, negotiated by the Commission
CART_2b	2nd stage of the Cartagena Protocol negotiations, negotiated by the Presidency
STOCPOP_1a	1st stage of the Stockholm Convention negotiations, negotiated by the Commission
STOCPOP_1b	1st stage of the Stockholm Convention negotiations, negotiated by the Presidency
STOCPOP_2a	2nd stage of the Stockholm Convention negotiations, negotiated by the Commission
STOCPOP_2b	2nd stage of the Stockholm Convention negotiations, negotiated by lead countries
SEA_1	1st stage of the SEA Protocol negotiations
SEA_2a	2nd stage of the SEA Protocol negotiations, negotiated by the Commission
SEA_2b	2nd stage of the SEA Protocol negotiations, negotiated by the Presidency

2.2. Qualitative Comparative Analysis

Analysing 18 cases confronts me with an 'intermediate-N situation', i.e. a situation with approximately 5 to 50 cases (Rihoux, Ragin, 2004). QCA is a suitable data-analysis method here, as it is appropriate to compare a limited number of cases in a systematic way. The idea behind QCA is developed by Charles Ragin (Ragin, 1987). He intended QCA as a bridge between qualitative (case oriented) and quantitative (variable oriented) approaches. More specifically, QCA aims to combine the advantages of both approaches: the empirical richness of every empirical case, and the analytical richness of formal examination (De Meur, Rihoux, 2002).

QCA uses a configurational logic, meaning that every case is considered as a particular configuration of conditions and an outcome (here: the EU negotiator's discretion). In other words, QCA uses a holistic view on cases: conditions have to be considered and assessed in the context of other relevant conditions. Cases are not seen as the aggregation of isolated variables, which may all have an independent effect on the outcome. This configurational logic is linked to QCA's 'multiple conjunctural causality' conception (Ragin, 1987). Applied to this research, it means, first, that a particular degree of discretion can be caused by a combination of conditions (*conjunctural*), and, second, that *multiple* (different) such combinations of conditions can cause that particular degree of discretion. A combination of conditions is called a 'causal path'.

Four stages can be distinguished in a QCA procedure: dichotomizating the raw data, minimizing the dichotomized data, obtaining a minimal formula, and interpreting the results in the light of the empirical data of the cases.

First, QCA requires dichotomized variables: each continuous variable needs to be recoded to '1' or '0', respectively indicating its presence (or high degree) or absence (or low degree). This is an important step in the procedure that has to be done rigorously and in interaction with the theories and the cases. However, dichotomizing the rich empirical data presented in the previous chapter is not an easy task because the variables do not have only two possible values by nature (absence or presence of a particular characteristic). As a result, each variable entails an interpretation: which cases constitute a subset of cases taking a lower value on a particular variable vis-à-vis a subset of cases taking a higher value on that variable? Once dichotomized, the values are then represented in a so-called 'truth table'. Dichotomization can be criticized because it decreases the information richness. However, the empirical nuances are reintroduced in the analysis as a final step of the QCA procedure, so information richness is not lost. Moreover, general tendencies can surface by looking at the broader picture (Rihoux, 2003). The strength of QCA is that rich empirical data can be interpreted through the lens of a framework, created by systematically compared cases.

Second, dichotomized variables allow executing the Boolean minimization procedure, which is at the heart of QCA. The purpose of this minimization process is to specify which combinations of conditions cause the outcome's absence or presence. The minimization procedure is based on pairwise comparisons of the configurations (cases): 'If two Boolean expressions differ in only one causal condition yet produce the same outcome, then the causal condition that distinguishes the two expressions can be considered irrelevant and can be removed to create a simpler, combined expression' (Ragin, 1987, p. 93). This formula has to be executed until further simplifications are no longer possible. An important principle of QCA is that minimizations have to be executed for both the 1 and the 0 outcomes because QCA does not assume symmetrical causality between 1 and 0.

Third, both the 0 and the 1 minimization lead to a minimal formula. This is a Boolean expression of the configuration of conditions that cause the outcome in question. Technically, a minimal formula is expressed as the sum of products, each of them being a causal path leading to a certain value on the outcome variable. A common practice in QCA literature is to present conditions with a 1 value in upper case and conditions with a 0 value in lower case. 'A' thus means 'the presence of variable A', while 'a' stands for 'the absence of variable A'. The Boolean operations 'and' and 'or' are respectively presented by '*' and '+'.

Fourth, the minimal formulae have to be interpreted in the light of the empirical data. This corresponds to the rationale behind QCA of going back and forth between the cases, the theory and the method. Moreover, a minimal formula only expresses which combinations of which conditions lead to a certain outcome, it does not say anything about the process and the causal mechanisms behind it (Rittberger, Schimmelfennig, 2005). In other words: the mechanisms behind the multiple conjunctural causation have to be discovered by reintroducing the empirical data of the cases into the analysis.

3. Step 1: dichotomizing the data

To dichotomize the empirical data presented in the various case studies in the previous chapter, I ranked, for each variable, the cases on a continuum going from low values to high values. It is impossible to define these absolute degrees of each variable for each case precisely on the basis of qualitative data collection as I opted for. However, comparing these values and determining their degrees *in relative terms*, i.e. by defining the value on a variable of one case *in relation to* the value on that variable of the other cases, can adequately be done on the basis of the available data. Then, a threshold was defined, creating subsets of cases for each variable. The main criterion is that these subsets have to be as homogeneous and as natural as possible (Cronqvist, 2003). This means that it is not the threshold's position *as such* that is significant, but the requirement that the threshold is located at a position where it accurately distinguishes cases with a lower degree from cases with a higher degree on a particular variable.

This way, all the condition variables could relatively easily be dichotomized. However, in order to explain the outcome variable 'discretion of the EU negotiator' in a highly qualified way, I opted to recode the outcome variable <disc> with three – instead of two – possible values: a low [0], moderate [1] or high [2] degree of discretion. Since QCA requires a dichotomized outcome variable, I then decoupled <disc> into two dummy variables: <disc_low> and <disc_high>. The dummy variable <disc_low> measures whether the case is characterized by a low degree of discretion (<disc_low> = 1) or by a moderate *or* high degree of discretion (<disc_low> = 0); <disc_high> measures whether the case is characterized by a high degree of discretion (<disc_high> = 1) or a moderate degree of discretion (<disc_high> = 0). Table 5.2 shows the values of the dichotomized dummy variables based on the multi-value outcome variable <disc>.

Table 5.2   Definition of the dummy outcome variables

Dummy variable	<disc>=0	<disc>=1	<disc>=2
<disc_low>=1 *if*	1	0	0
<disc_high>=1 *if*	–	0	1

As Table 5.2 reveals, the cases characterized by a low degree of discretion for the EU negotiator (<disc>=0) do not get a value on the dummy variable <disc_high>. These cases will thus not be included in the analysis and minimization procedure of <disc_high>, only in the minimization procedure of <disc_low>. As a result, the minimization procedures will be executed in two broad steps. First, the causal paths leading to the presence and absence of a low degree of discretion are identified (Section 4.1). Second, in this study, I want to go beyond explaining the occurrence and non-occurrence of discretion. This second step intends to explain under which conditions an EU negotiator enjoys a high degree of discretion (Section 4.2). Briefly, in step 1, I examine under which conditions an agent enjoys discretion. In step 2, I answer the question: if an agent enjoys discretion, how can the agent's *particular degree* of the discretion be explained?

The dichotomized data is presented in Table 5.3. The first eight columns provide the dichotomized values for the conditions. In the next columns, the multi-value outcome variable and the dummy outcome variables are presented, followed by the technical cases corresponding to the configurations.

Table 5.3   Truth table

prefprin	prefpa	polit	infag	infprin	extcomp	agprin	instdens	disc	disc_low	disc_high	Cases
0	0	0	0	0	0	0	0	0	1	-	/ARHUS_1(prefpa:0)(agprin:0)/
0	0	0	0	0	0	1	0	0	1	-	/ARHUS_1(prefpa:0)(agprin:1)/
0	1	0	0	0	0	0	0	0	1	-	AEWA; /ARHUS_1(prefpa:1)(agprin:0)/; ARHUS_2; SEA_1; SEA_2a;
0	1	0	0	0	0	1	0	0	1	-	/ARHUS_1(prefpa:1)(agprin:1)/; SEA_2b
1	1	0	0	0	0	1	0	0	1	-	CCD
0	1	0	0	0	1	0	0	1	0	0	CART_1a
0	1	0	0	0	1	1	0	1	0	0	CART_1b
1	1	1	0	0	1	1	1	1	0	0	CART_2b; STOCPOP_1b; STOCPOP_2b
1	1	1	0	0	1	0	1	1	0	0	STOCPOP_1a; STOCPOP_2a
0	0	1	0	1	1	1	1	2	0	1	KYOTO_1
0	0	1	1	1	1	1	0	2	0	1	KYOTO_2
1	1	0	0	0	1	0	1	2	0	1	PIC
1	1	1	1	0	1	0	1	2	0	1	CART_2a

4. Steps 2 and 3: minimizing low, moderate and high degrees of discretion, and calculating the minimal formulae

4.1. Explaining a low degree of discretion: minimizing <disc_low>

By minimizing the 1 and 0 values of <disc_low>, I examine respectively the causal paths leading to a low degree of discretion and the causal paths leading to a non-low degree (i.e. a moderate *or* a high degree) of discretion. As a consequence, by minimizing the 0 outcomes of <disc_low>, I am able to discover the paths leading to cases in which the negotiator enjoyed a *certain degree* of discretion, without being able to say something about the particular degree (is it moderate or high?).

The minimization of <disc_low>=1 results in the following minimal formula[1]:

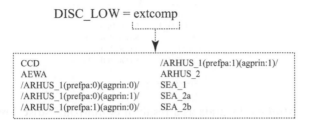

$$DISC\_LOW = extcomp$$

CCD	/ARHUS_1(prefpa:1)(agprin:1)/
AEWA	ARHUS_2
/ARHUS_1(prefpa:0)(agprin:0)/	SEA_1
/ARHUS_1(prefpa:0)(agprin:1)/	SEA_2a
/ARHUS_1(prefpa:1)(agprin:0)/	SEA_2b

Figure 5.1   Minimal formula as a result of minimizing <disc_low>=1

According to this formula, a low degree of discretion can be explained by a single condition: the absence of a compelling external environment. This means that a low degree of external compellingness is a sufficient and a necessary condition for a low degree of discretion. Or in other words: a non-compelling international context is enough to observe a low degree of discretion, irrespective the preference distances, information benefits, level of politicization, institutional density or the affiliation of the agent.

To explain the absence of a low degree of discretion (i.e. the presence of some discretion, without being able to judge the particular degree), <disc_low>=0 is now minimized:

---

1   One of QCA's best practices is to execute the minimizations of the 0 and 1 outcomes, both with and without the inclusion of so-called 'remainders' (De Meur, Rihoux, 2002; Rihoux, Ragin, 2004). Remainders are configurations of variables with dichotomous values, which do not correspond to a case. The advantage of including remainders in the analysis is that the results become more parsimonious and more explanatory powerful. In this study, the minimization procedures without the inclusion of the remainders only generated minimal formulae which mostly came down to the sum of the various minimized configurations and which were thus not parsimonious. They are not presented here. By contrast, the minimizations with the inclusion of the remainders, being more parsimonious, offer more analytical power to understand the EU negotiator's degree of discretion. The inclusion of the remainders can be justified by two additional reasons. First, the ratio of the number of conditions to the number of cases is quite large. Second, the cases for this research are selected in such a way that all recent EU decision-making processes with regard to an international negotiation leading to an MEA, which is signed by the EC and the member states (according to my definitions, see Chapter 3) are included in the analysis. As a result, I am sure that I do not include remainders, of which the configuration corresponds to an empirical case with the opposite outcome than the minimized outcome for which the remainder is used.

disc_low = EXTCOMP

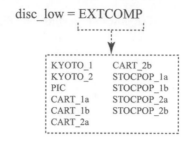

KYOTO_1	CART_2b
KYOTO_2	STOCPOP_1a
PIC	STOCPOP_1b
CART_1a	STOCPOP_2a
CART_1b	STOCPOP_2b
CART_2a	

Figure 5.2   Minimal formula as a result of minimizing <disc_low>=0

This minimal formula shows that the absence of a low degree of discretion goes together with the presence of a compelling external environment, which is thus a necessary and sufficient condition for a moderate or high degree of discretion. When the external environment is compelling, the agent enjoys a degree of discretion that is higher than what I have labelled as a 'low degree of discretion'.

As mentioned above, QCA does not automatically imply symmetrical causality. However, in this case, the minimization of the 0 and 1 outcomes of <disc_low> generates symmetrical results. A compelling external environment leads to an at least moderate degree of discretion, while a non-compelling external environment causes a low degree of negotiation discretion for the EU negotiator. This means that, for the explanation of a low degree of discretion, symmetrical causality is at stake.

4.2. Explaining a high and moderate degree of discretion: minimizing <disc_high>

On the basis of the findings of the previous section, I am only able to conclude that the degree of compellingness of the external environment is decisive for the absence or presence of a low degree of discretion: when the external environment is not compelling, the EU negotiator's discretion is low; when it is compelling, the discretion is not low. However, it is not possible yet to assess how much discretion the agent enjoys when an EU decision-making process is characterized by a compelling external environment. Therefore, I now examine those cases that are characterized by a non-low (i.e. a moderate or a high degree) degree of discretion in order to verify how much discretion the agent enjoyed in those decision-making processes. As has become clear from the minimization of <disc_low>, these cases are all characterized by a compelling external environment. Since I made a distinction between cases with a moderate degree of discretion (<disc>=1) and cases with a high degree of discretion (<disc>=2), I am not only able to analyse the causal paths leading to some discretion, but also the causal path(s) leading to a particular non-low (i.e. moderate or high) degree of discretion. To explain the presence and absence of a high degree of discretion for the agent, the dummy outcome variable <disc_high> is minimized.

The minimization of <disc_high>=1,R resulted in the following minimal formula[2]:

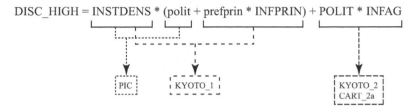

Figure 5.3   Minimal formula as a result of minimizing <disc_high>=1

This minimal formula shows that three causal paths lead to a high degree of discretion:

- [PATH 1 AND 2] a decision-making process that is institutionally dense *and*
  - [PATH 1] *either* not politicized (like *PIC*);
  - [PATH 2] *or* characterized by heterogeneous preferences among the principals and an information benefit for the principals (like *KYOTO_1*);
- [PATH 3] *or* a decision-making process that is
  - politicized; *and*
  - in which the agent has an information advantage vis-à-vis the principals (like *KYOTO_2* and *CART_2a*).

To explain the absence of a high degree of discretion (and thus the presence of a moderate degree of discretion), <disc_high>=0 is minimized:

---

2   For the minimization of <disc_high>, two manual interventions were needed in the QCA procedure. First, I opted to select the so-called 'prime implicants' manually. It is also possible to allow for an automatic selection of the prime implicants by the software, but doing this selection manually allows the researcher to control the procedure and to ensure that the choices that are made are both theoretically and empirically justifiable. Second, the inclusion of the remainders generates the risk that so-called 'contradictory simplifying assumptions' are used in the analysis. Therefore, I checked for these contradictory simplifying assumptions, and when these were indeed used, I solved them (Delreux, Hesters, 2010). This happened twice, as a result of which the contradictory simplifying assumptions were solved twice as well. Also this manual solution of contradictory simplifying assumptions was driven by a double consideration: theoretical and empirical consistency. For reasons of transparency and replicability of this study, I am happy to provide a logbook of these various manual steps to the reader who is interested in it.

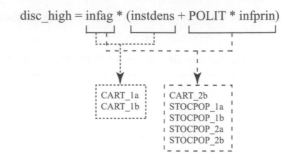

Figure 5.4   Minimal formula as a result of minimizing <disc_high>=0

This minimal formula shows that two causal paths lead to the absence of a high degree of discretion:

- [PATH 1] the combination of
    - the absence of private information for the agent; *and*
    - the absence of an institutionally dense decision-making setting;
- [PATH 2] *or* the combination of
    - the absence of private information for the agent;
    - the presence of a level degree of politicization; *and*
    - the absence of private information for the principals.

The absence of an information benefit for the EU negotiator is a necessary condition for a moderate (i.e. non-high) degree of discretion. This means that, if the agent does not have private information, he cannot enjoy a high degree of discretion.

5. Step 4: interpreting the minimal formulae by going back to the cases

In this section, I reintroduce the empirical data and the case studies in the analysis. Starting from the minimal formulae, I return to the data that is needed to explain a low and a high degree of discretion. The detailed and in-depth case study material can of course be found in the eight descriptions of the cases in the previous chapter. For the explanation and interpretation of a low, high and moderate degree of discretion, I firstly focus on the minimal formulae (first the outcome/discretion side and then the side of the conditions and the causal path(s) of the formula), after which I will touch upon the causal mechanism and the causal link between both sides. The results of the QCA procedure only point out which combinations of which conditions lead to a particular outcome, but a full understanding of discretion also requires to specify *why* a causal path causes a particular outcome.

5.1. Decision-making processes with an agent enjoying a (non-)low degree of discretion

5.1.1. Minimal formulae and the empirical data

In four EU decision-making processes, corresponding to six technical cases, the agent enjoyed a low degree of discretion: the decision-making processes with regard to the UNCCD, AEWA, the Aarhus Convention and the SEA Protocol. The EU negotiator had a very limited leeway vis-à-vis the member states, due to a low degree of both granted and conquered autonomy.

In the first eight stages of the Aarhus Convention negotiations (*ARHUS_1*), there was no EU negotiator, and thus no discretion, as the Commission's negotiation authority was formally contested. The member states did not grant an authorization and a mandate to the Commission for the greatest part of the negotiations because of the uncertainty about the status of the Community institutions under the future Aarhus Convention provisions. Moreover, some member states[3] wanted to retain the possibility of expressing their own preference at the international level. Also in (the beginning of) the negotiations leading to the SEA Protocol, the negotiation authority of the Commission was, although informally, contested. Unlike in the decision-making process with regard to the Aarhus Convention negotiations, the Commission was formally granted a mandate, but the member states did not maintain this EU negotiation arrangement with the Commission as their exclusive representative. Particularly in the first stage of this negotiation process (*SEA_1*), the EU negotiation arrangement was very unstructured, with both the principals and the (formally appointed) agents taking the floor. Moreover, it was not always clear on whose behalf member states and the Commission were speaking (on their own behalf or on the behalf of the EU as a whole?). In the negotiations on both the Aarhus Convention and the SEA Protocol, the member states mostly negotiated in their capacity as UNECE member state. From an EU point of view, these negotiations were intergovernmental in nature.

In the negotiations leading to UNCCD and AEWA, the negotiation authority of the agent, respectively the Presidency and the Commission, was not contested by the member states. However, like it was the case during the Aarhus Convention and SEA Protocol negotiations, the member states could take the floor and express their own preferences. On most of the issues of these two decision-making processes, the member states agreed on the common European line (no new financial mechanism for UNCCD, not going beyond the Bird Directive for AEWA). However, when a member state had a diverging preference on a particular issue, it was allowed to express that preference at the international level. Moreover, the representation role of the Presidency decreased when the UNCCD negotiations evolved, as the agent role was *de facto* taken over by the Australian WEOG Presidency of that time.

---

3   More in particular, member states like Italy and Denmark, which had maximalist preferences.

The decision-making processes in which the agent enjoyed a low degree of discretion were characterized by the possibility for the member states to attend all international negotiation settings. Furthermore, frequent coordination meetings took place, resulting in strict instructions from the principals to the agent. However, two exceptions should be mentioned. First, in the first stage of the Aarhus Convention negotiations (*ARHUS_1*), EU coordination meetings were only seldom organized (mostly once each negotiation session), which was due to the absence of an EU negotiator and the fact that the negotiations were *de facto* not conducted via the EU level. Second, in the negotiations on UNCCD, EU coordination meetings between the agent and the principals were organized, but they were supplemented by WEOG coordination meetings, which were considered even more important than the EU coordinations. (This can be explained by the North-South division in the INCDs.) This led to a situation where the forum of the developed (WEOG) countries was used as the main coordination mechanism. The less frequent and less intensive EU coordination meetings during the Aarhus Convention and the UNCCD negotiations can thus not be seen as an indication of a higher degree of autonomy granted to the agent.

Finally, in none of the EU decision-making processes with an EU negotiator enjoying a low degree of discretion did the agent conquer additional autonomy. I have no indications that an EU negotiator went beyond the instructions of the member states or that agency slack occurred.

The result of the QCA procedure shows that the agent's low degree of discretion, as illustrated above, is determined by the lack of a compelling external environment. The negotiations resulting in AEWA, the Aarhus Convention and the SEA Protocol were the only negotiations studied of which the scope is not global, but regional or intercontinental. In these negotiation processes, the relative bargaining power of the EU was large. Such international negotiations generate less compellingness on the principals not to jeopardize the negotiations than global, UN- wide, negotiations do. Some decision-makers even admitted that the EU *de facto* 'dominated' these negotiations. AEWA was negotiated with 22 African and 19 Central and Eastern European (and even 'Eurasian') countries, which could not deploy a large bargaining power vis-à-vis the EU in these negotiations. The negotiations on the Aarhus Convention and the SEA Protocol took place under the auspices of UNECE, a negotiation forum in which the EU was the largest and strongest player. In practice, the negotiation partners that were able to play a significant role in the Aarhus Convention negotiations were Russia, a couple of Central and Eastern European countries, Norway and the NGOs. In case of the SEA Protocol negotiations, this group was yet more restricted: Switzerland, Norway, the Espoo Convention Secretariat[4] and the NGOs. The at that time candidate member states were also attending the EU coordination meetings, which made their possible opposition to the EU at the UNECE level less likely. Hence,

---

4   The mother treaty of the SEA Protocol is the Espoo Convention, which explains the role for the latter's secretariat in the negotiation process leading to the SEA Protocol.

the external compellingness in the EU decision-making processes with regard to the international negotiations leading to AEWA, the Aarhus Convention and the SEA Protocol was limited because of the non-global scope of these negotiations. This was, furthermore, strengthened by the fact that the issues at stake did not deal with urgent environmental problems or threats, but with procedures to enhance environmental decision-making (Aarhus Convention and SEA Protocol) or with exporting the EU's way of nature conservation outside the EU borders (AEWA).

Although the negotiations leading to UNCCD had a global scope, their compellingness was limited as well. The EU member states were not the demanding party for a multilateral convention on desertification. Such a multilateral instrument to tackle the desertification problem was considered neither important for the EU players nor the appropriate tool to solve the issue. The only factor of compellingness experienced in the EU was the risk that a failure of the negotiations would have a negative political spill-over effect on other multilateral environmental governance processes (in particular on climate change and biodiversity).

5.1.2. Causal mechanisms
What is the link between the external compellingness of a particular EU decision-making process on the one hand, and the degree of discretion of the EU negotiator in that decision-making process on the other hand? Why does an EU negotiator enjoy a *low degree of discretion* in international negotiations with a low degree of external compellingness? There are two causal mechanisms functioning behind this causal relation: neither by the agent nor by the principals, is discretion really needed to reach an agreement at the international level.

First, there is only a limited number of negotiation partners and they have relatively little bargaining power. In such situation, the member states can permit themselves to undermine the bargaining-power benefit of delegation, e.g. by taking the floor themselves. In an EU decision-making process with a low degree of external compellingness, the incentives for the principals to grant a high degree of autonomy to the agent are limited and they can still deploy their control mechanisms without putting on the line the effectiveness of the EU at the international level. In negotiations where the EU is confronted with relatively weak negotiation partners, member states are aware that the decision-making process at the EU level becomes more and more important for influencing the outcome of the negotiations than the process at the international level. If the relative bargaining power of the EU is large, the outcome of the negotiations is likely to be largely determined by the EU, and thus by the decision-making process at the EU level. As a result, principals make use of the control mechanisms at the EU level, which implies controlling the agent, and thus a lower degree of discretion.

Second, it is unlikely that the EU negotiator in a non-compelling context will conquer more autonomy than granted by the member states. The main reason (and justification) for an EU negotiator for conquering more autonomy is exactly the pressure coming from the external negotiation partners to move without the principals' consent. If the bargaining power of the negotiation partners is limited, they cannot force the EU negotiator to make concessions the member states have not previously agreed with. In such a situation, the agent cannot exploit any external compellingness in his relation with the principals.

The causal mechanism between a high degree of external compellingness and *a moderate or high degree of discretion* for the EU negotiator is more clear and less surprising. It confirms the dynamics of my first extension of the principal–agent model (see Section 3.1 of Chapter 3, and in particular Figure 3.4): a compelling external environment, characterized by a large relative bargaining power for the negotiation partners vis-à-vis the EU, leads to a high political cost of no agreement for the member states. In such situation, their final assessment of the MEA will be influenced by, on the one hand, the substantive costs and benefits of the provisions of the MEA in relation to their initial preferences and, on the other hand, the additional political costs related to a (political or legal) rejection of the MEA. As this cost of no agreement is high when the external environment is compelling, the final win-set of the principals expands (Delreux, 2008). The key element in explaining discretion is that the agent anticipates on this dynamic. In such a situation, the agent can exploit his discretion because he has an instrument to do so: the compellingness of the external environment. The EU negotiator is able to negotiate with a higher degree of discretion since he is aware that the likelihood of being called back by the member states is low in such situation.

5.2. Decision-making processes with an agent enjoying a high degree of discretion

In four technical cases, the agent enjoyed a high degree of discretion: the Commission in the negotiations leading to the PIC Convention (*PIC*), the Presidency and the troika during the Kyoto Protocol negotiations (*KYOTO_1* and *KYOTO_2*) and the Commission in the second half of the negotiations that resulted in the Cartagena Protocol (*CART_2a*). As the QCA results show, three causal paths lead to a high degree of discretion for the agent: the combination of an institutionally dense and non-politicized decision-making process, the combination of a high degree of institutional density, heterogeneous preferences among the member states and an information benefit for them, or the combination of a politicized decision-making process in which the agent has private information. The first path covers *PIC*, the second path *KYOTO_1*, and the third path *KYOTO_2* and *CART_2a*.

5.2.1. Causal path 1: INSTDENS * polit

5.2.1.1. Minimal formula and the empirical data   The high degree of discretion enjoyed by the Commission in the PIC Convention negotiations revealed itself in the fact that the Commission was not only the EU negotiator for the issues covered by EC competences, *in casu* by Regulation 92/2445, but also for the other issues. The member states did not question the negotiation authority of the Commission and they even granted more autonomy, as a result of which the Commission was *de facto* the only agent. Moreover, the Commission was able to negotiate on the basis on of a high degree of granted autonomy. The basic instruction of the member states – corresponding to the preferences of the Commission – was that the provisions of the PIC Convention could not be incompatible with the existing EC Regulation. However, within the boundaries of this instruction, the Commission could act with a high degree of autonomy. This high degree of discretion was not undone by the facts that the member states attended all the international negotiation settings, that EU coordination meetings were regularly organized, or that the agent did not conquer additional autonomy.

The Commission's high degree of discretion during the PIC Convention negotiations can be explained by the lack of political sensitivity on the PIC issue, combined with a highly institutionalized decision-making context, which made the decision-making process between the member states and the negotiating Commission more cooperative than one would expect from a traditional RCI perspective.

This high degree of institutional density in the EU is linked to the fact that the officials from the member states and the Commission were already engaged in EU decision-making on European chemicals policy during many years. Most member state representatives came from their Designated National Authority, responsible for the implementation of Regulation 92/2445. In this capacity, they regularly met in expert committees in Brussels, together with the Commission officials. This contributes to the understanding why all the institutional norms of the Council that I distinguished (institutional memory and diffuse reciprocity, trust and credibility, mutual responsiveness, consensus and compromise striving, and avoiding process failure) were present during the decision-making process with regard to the PIC Convention. Especially the large degree of trust in the Commission representatives, built during the years before the PIC Convention negotiations started, generated a good team spirit in the EU.

Such a dense institutional environment was complemented by the absence of political sensitivity for the PIC issue in the member states. Since the PIC procedure was regulated in the EC Regulation mentioned above, the negotiations were not politicized at all. Moreover, the decision-making process was conducted by technical experts and there was little NGO or media attention for the negotiations.

5.2.1.2. Causal mechanism In a highly institutionalized and non-politicized decision-making process, the risk of granting a high degree of autonomy to the agent is not high for the principals. As member state and Commission officials engage in a socialization process, which is facilitated by the existing relations of personal trust and by a shared history of common decision-making (*in casu* on chemicals regulations), the likelihood of agency slack is very limited. The reason is twofold. On the one hand, in such a situation, the principals trust the agent, as a result of which they do not tie the EU negotiator's hands and they do not fully deploy the control mechanisms. On the other hand, even if there remains a risk of agency slack (or too much discretion, from the perspective of the principals) because of the compelling external environment, member states are likely to go along with the reasoning and the negotiation behaviour of the agent. They will accept the agent's potential to conquer more autonomy because the institutionally dense relation between principals and agent generates a kind of empathy by the principals for the fact that the agent is confronted with a compelling external environment. Moreover, like this causal path indicates, there is little risk that the actors at the various domestic levels will be highly concerned about the EU decision-making process, the international negotiations and its outcome, because the issues at stake are not politically sensitive in the member states. If there are little sensitivities at stake, the agent cannot do much wrong.

This first causal path thus relates to a situation in which the EU decision-making process develops, on the one hand, rather freely from the domestic levels, and, on the other hand, in a mutual cooperative way. This leads to a situation in which the member states grant a large degree of autonomy to the agent in order to realize their compromise at the international level. Hence, in such a decision-making process, in which national representatives negotiate politically insensitive issues in a cooperative atmosphere, the EU negotiator enjoys a high degree of discretion.

5.2.2. Causal path 2: INSTDENS * prefprin * INFPRIN

5.2.2.1. Minimal formula and the empirical data   During the first half of the Kyoto Protocol negotiations, i.e. during the first six months of 1997 (*KYOTO_1*), the Dutch Presidency enjoyed a high degree of discretion. This did not so much reveal itself in the negotiation behaviour of the agent, but certainly in its agenda-setting role in the 1997 Burden Sharing Agreement negotiations. The Dutch Presidency claimed and realized ownership on the climate change issue (e.g. by executing studies, presenting simulations and organizing seminars). By putting forward different burden-sharing proposals in the Council, the Dutch Presidency enjoyed an autonomous agenda-setting power. This allowed the agent to act with a high degree of discretion at both the international and the EU level (for the agenda-setting). Moreover, in the first months of 1997, the Dutch Presidency was in any case considered a strong actor with a large degree of action capacity. The regular EU coordination meetings, the attendance of the member states in the Subsidiary Bodies and the strict Council conclusions of March 1997 did not undermine the EU negotiator's high degree of discretion.

The high degree of discretion of the Dutch Presidency can be explained by the institutional density of the decision-making process, combined with the heterogeneous preferences of the member states, which did not reveal their fallback positions. The preferences of the member states on the national emission reduction targets considerably differed in the first half of 1997. The decision-making process at the EU level is regarded as very tough, because of the heterogeneous preferences among the member states. Maximalist member states like Austria, Sweden, Denmark and Germany strived for strong emission-reduction targets, while minimalist principals like Spain, Portugal, Greece and Italy were not willing to impose themselves with strong reductions, which would hinder their economic growth. The member states did not only have diverging preferences, neither were they open about their fallback positions vis-à-vis the agent. Some member states were presumably able to accept stronger reduction targets. This means that they had private information about their fallback positions and that they managed to keep this information asymmetrically divided. This constellation of preferences (diverging and not completely transparent) took place in a densely institutionalized decision-making setting. In the first stage of the Kyoto Protocol negotiations, institutional norms played a much larger role in the EU than in the second stage (*KYOTO_2*). The understanding of each other's domestic problems on national reduction targets and the role of 'fairness' (Chagas, 2003) in the 1997 Burden Sharing Agreement negotiations made the institutional decision-making context quite dense (Vogler, 2009). Moreover, member states knew that they were dependent on each other for the realization of their common strategy, i.e. to get the bubble concept included in the international agreement in order to regulate the intra-EU division of reduction targets via the internal Burden Sharing Agreement, which was politically already agreed upon in March 1997.

5.2.2.2. Causal mechanism   From a theoretical point of view, it is not surprising that the combination of an institutionally dense setting, member states not putting all their cards on the table and having heterogeneous preferences leads to an agent enjoying a high degree of discretion in a decision-making process with a compelling external environment.

   A high degree of institutional density, and in particular a high extent of trust of the principals in their agent, reduces the extent to which the member states deploy the control mechanisms because they do not assume that the agent will negotiate against their wishes, and thus conquer additional autonomy. Moreover, as the principals and the agent are engaged in a dense partnership atmosphere, the principals show a kind of empathy for the fact that the agent is facing a compelling external environment and the pressure to make concessions at the international level. In case of *KYOTO_1*, this trust and empathy were combined with heterogeneous preferences among the member states, which allowed the Dutch Presidency to play the member states against each other in its proposals of the 1997 Burden Sharing Agreement. Indeed, being faced with diverging member state preferences, there was no risk that these proposals would be rejected by the whole group of principals. Some member states would oppose it and others would support it, which creates bargaining opportunities for the agent. Given the institutionally dense environment, this even extended the Presidency's discretion. Moreover, the discretion got an additional increase because the agent was aware of the fallback positions of the principals. This confirms my first extension of the principal–agent model, which presumes that the existence of private information of the principals and the anticipation of the agent on this asymmetrically divided information is likely to increase the latter's discretion. The reason is that, in such situation, the EU negotiator can anticipate on the high political cost of no agreement for the member states, which broadens their final win-set. In a decision-making process like *KYOTO_1*, the agent can exploit the external compellingness to force the principals to ultimately accept more than they have hinted at during the process. Briefly, the causal mechanism behind this causal path comes to a mixture of conditions of which it is theoretically plausible to expect such an effect, and which combination also generates that effect in practice.

5.2.3. Causal path 3: POLIT * INFAG

5.2.3.1. Minimal formula and the empirical data   The final stages of the Kyoto and Cartagena Protocol negotiations (*KYOTO_2* and *CART_2a*) were characterized by a high degree of discretion for, respectively, the troika and the Commission. In both decision-making processes, the leeway for the agent was considerably broadened at the end to reach an international agreement. In Kyoto, and mainly in the hectic last days, the negotiating troika could negotiate without being strictly bound by the principals. These intense final negotiations took place in a Friends of the Chair setting on the one hand, and in a group with the EU, the US and Japan on the other hand. In both settings, the EU was only represented by the troika, without the member states attending these meetings. The member states were no longer able to give detailed instructions to their negotiator, despite the quasi-continuous EU coordination meetings. Moreover, the troika occasionally conquered more autonomy than the autonomy that was granted by the member states. Indeed, *KYOTO_2* is the only case studied where agency slack – the agent acting against the wishes of the principals – occurred. In the endgame of the negotiations, the troika accepted some proposals without the backing of the member states. These proposals were even opposite to the preferences of some of them. The troika is said to have given in too quickly on the -15% reduction position of the EU, and the UK (being one of the three troika members) seemed to have made commitments on behalf of the EU about the sinks issue without the support of the member states.

The Commission also enjoyed a high degree of discretion in the second stage of the Cartagena Protocol negotiations. After BSWG 6, the Commission became *de facto* the only EU negotiator, as the agent role of the Presidency decreased. Moreover, the Commission occupied the only EU seat in the Group of Ten setting, which was used in the ExCOP in Cartagena. Furthermore, the Council conclusions of December 1999 about the final negotiation session of the Cartagena Protocol negotiations (Council of Ministers, 13854/99) granted a high degree of autonomy to the EU negotiator. These conclusions, being the formal instructions from the principals to the agent, emphasized the importance of reaching *an* agreement in Montreal (ExCOP-bis).

The high degree of discretion in *KYOTO_2* and *CART_2* can be explained by the combination of a highly politicized decision-making process in which the agent enjoyed an information benefit vis-à-vis the principals. The climate change and the GMO issue were politically very sensitive in the final stage of the negotiation processes. The whole world was following the Kyoto negotiations, which was going to result in a Protocol with an enormous impact on various domestic policies of the parties to the Protocol. Likewise, the GMO issue had reached the top of the political agenda in Europe at the end of the 1990s. Resistance in the public opinion against GMOs, triggered by food crises, NGO campaigns and green parties in government, made the final stage of the Cartagena Protocol negotiations very politicized in the EU member states. The fact that both negotiations were conducted at ministerial level illustrates their high level of political sensitivity.[5]

In both cases, the agent had an information benefit. In the endgame of the Kyoto negotiations, the continuous Friend of the Chair meetings and negotiations with the US and Japan, were only attended by the troika and not by the principals. The member states could not be kept informed about these crucial negotiations, despite the frequent EU coordination meetings. However, member state officials admitted that the agent did not strategically withheld information vis-à-vis the principals. The asymmetrical division of information about the substance and the developments of the endgame of these negotiations can be attributed to the intensity of the international negotiation process. During the second stage of the Cartagena Protocol negotiations, the agent's information benefit was not as large as it was for the troika in Kyoto. However, in the Group of Ten negotiation setting, only the Commission participated. This created a situation of asymmetrically divided information in favour of the EU negotiator. When this negotiation setting was not used anymore in the final negotiation sessions, the information about the international negotiation process became again more symmetrically divided between the agent and the principals. However, because of his central role, the agent maintained an information benefit.

5.2.3.2. Causal mechanism   The combination of a high level of politicization and an information benefit for the EU negotiator coincides with the occurrence of restricted negotiation settings at the end of the negotiations. In such settings with limited participation, the EU is only represented by the agent, and the principals are not able to attend. This dynamic of making the final deal in a Friends of the Chair (alike) negotiation group is mostly used in international negotiations dealing with highly politicized issues (like climate change and GMOs). Those issues are of course not only politically sensitive for the EU, but also for (most of) the other major negotiation partners. To cut the final – and almost *a priori* thus most politicized – knot, a limited negotiation setting with the key players often seems to be the only solution to reach an international agreement. Such a setting then mostly generates an information benefit for the agent, as the principals are *de facto* disconnected from the international negotiation process and as there no time left to consult the principals. This considerably reinforces the position of the EU negotiator, which

---

5   Moreover, in case of the EU decision-making process with regard to the Kyoto Protocol negotiations, a number of European environment ministers had invested a lot of political capital in the success of global climate change negotiations in the 1990s.

makes an increasing degree of discretion very likely. Hence, in a decision-making process characterized by a high degree of compellingness, the agent can optimally exploit his Janus-like role. The agent experiences two kinds of pressure. First, as only cases with a high degree of external compellingness are analysed here, the agent is confronted with pressure from the negotiation partners at the international level. Second, as the level of politicization is high, the agent also experiences pressure from the principals at the EU level. Because of this Janus-like role on the one hand and the strategic information benefit for the agent on the other hand, the agent can exploit the external compellingness towards the principals, as a result of which he is able to preserve his discretion. Hence, the combination of a high degree of politicization and asymmetrically divided information in favour of the agent generates a high degree of discretion in a decision-making process characterized by a high degree of external compellingness.

5.3. Decision-making processes with an agent enjoying a moderate degree of discretion

As mentioned above, the minimization of <disc_high> is only performed with cases that are characterized by a certain (i.e. a moderate *or* a high) degree of discretion. This means that the cases with a 0 value on <disc_high> are the cases with a 1 value on <disc> ('moderate degree of discretion'). Consequently, the minimal formula derived from the minimization of <disc_high>=0 explains the presence of a moderate degree of discretion for the EU negotiator in international negotiations leading to an MEA. By interpreting this minimal formula in the light of the empirical data, the explanation of this particular degree of discretion can be fully understood. There are two causal paths leading to a moderate degree of discretion: the combination of the absence of an information benefit for the agent and the absence of an institutionally dense decision-making context (covering *CART_1a* and *CART_1b*) on the one hand, and the combination of the absence of information benefit for both the agent and the principals and the presence of a highly politicized decision-making process (covering *CART_2b*, *STOCPOP_1a*, *STOCPOP_1b*, *STOCPOP_2a* and *STOCPOP_2b*) on the other hand.

5.3.1. Causal path 1: infag * instdens

5.3.1.1. Minimal formula and the empirical data In the first stages of the negotiations leading to the Cartagena Protocol (i.e. from BSWG 1 to 6), both the Commission (in *CART_1a*) and the Presidency (in *CART_1b*) enjoyed a moderate degree of discretion. After the contestation by the member states of the Commission's negotiation authority in BSWG 1, a gentlemen's agreement was agreed upon about the division of labour between the two agents: the Commission and the Presidency. From that moment on, the EU negotiators enjoyed discretion, but not a high degree. The Presidency and the Commission took the exclusive representation role, but their discretion could not increase to a high degree because three control mechanisms functioned effectively: the agents had to negotiate on the basis of detailed instructions by the principals (in the form of position papers, including the EU's fallback positions and the room of manoeuvre for the agent), the principals were able to attend all international negotiation settings, and the negotiation stage of the decision-making process was characterized by a continuous feedback from the agents to the principals, both in frequently organized coordination meetings and with the member states that attended the international negotiations.

This moderate degree of discretion can be explained by the combination of the absence of an information benefit for the EU negotiator and the absence of a high degree of institutional density. Officials from both the agent and the principal side confirm that the flow of information from the Commission and the Presidency to the member states was optimal and that the member states could be perfectly informed about the substance and the developments of the international negotiations. Moreover, the member states were able to attend all international negotiation settings, which allowed them to rely on first-hand information. These decision-making processes were not characterized by a high degree of institutional density. None of the five institutional norms played a significant role from BSWG 1 to 6. This changed completely in the second stage of the decision-making process (*CART_2a* and *CART_2b*).

5.3.1.2. Causal mechanism In a decision-making process characterized by a compelling external environment, the EU negotiator can exploit this compellingness, which gives him discretion vis-à-vis the member states. Indeed, in such a situation, the agent is able to play his role as EU negotiator in an exclusive way, as the principals do not undermine the agent's representation role (e.g. by taking the floor themselves). However, in a decision-making process in which the agent has no information benefit and in which the degree of institutional density is not high, an agent is not able to extend its discretion to a high degree. On the one hand, as the information about the substance and the developments of the international negotiations is symmetrically divided between the agent and the principals, the agent has no strategic information benefit in his relation with the principals. On the other hand, as the degree of institutional density is rather low, the member states do not trust their negotiator and they do not show empathy for the (compelling) negotiation situation with which the EU negotiator is confronted at the international level. More in particular, the fact that the relation between the principals and the agent is not very cooperative prevents that the member states empathically go along with the negotiation behaviour of the EU negotiator, who experiences the compellingness of the external environment to make concessions towards the external partners.

*5.3.2. Causal path 2: POLIT * infag * infprin*

5.3.2.1. Minimal formula and the empirical data The moderate degree of discretion enjoyed by the Presidency in *CART_2b* and by the Presidency, the Commission and the lead countries in the two stages of EU decision-making process with regard to the Stockholm Convention negotiations (*STOCPOP_1a*, *STOCPOP_1b*, *STOCPOP_2a* and *STOCPOP_2b*), has the same characteristics as the agent's moderate degree of discretion during *CART_1a* and *CART_1b* that I described in Section 5.3.1.1: the agents were the only EU actors who took the floor, but they acted under clear instructions of the principals (issued at ministerial level in the form of Council conclusions). Moreover, the member states could attend all international negotiations, there were frequent EU coordination meetings and there were no indications of an agent conquering more autonomy than granted by the member states.

This moderate degree of discretion can be explained by a politicized decision-making context, in which neither the agent nor the principals enjoyed an information benefit. The precautionary principle was the main issue that made both the Stockholm Convention negotiations and the second stage of the Cartagena Protocol negotiations highly politicized. This principle took a very high position on the political agenda in Europe by the end of the 1990s. In the final stages of both negotiations, the inclusion of the precautionary principle in the MEA had become the main priority for the EU. In addition to this, the political sensitivity of the EU decision-making process with regard to the second stage of the Cartagena Protocol negotiations (i.e. from the ExCOP in Cartagena to the ExCOP-bis in Montreal) was also caused by an increasing resistance of the public opinion and NGOs against GMOs, the entry of green parties in a number of European governments, increasing consumer concern about food safety and the simultaneous discussions about the *de facto* moratorium on GMOs in the EU. This high level of politicization is not sufficient to a moderate degree of discretion: it needs to be combined with a symmetrical division of information between principals and agents, both about the substance and developments of the international negotiations and about the preferences and fallback positions of the principals. Since there were no negotiation settings that could not be attended by the member states, and since representatives from the principals and from the agents acknowledge that the flow of information from the EU negotiator to the member states was optimal, the agent did not enjoy an information benefit. Likewise, both the Presidency in *CART_2b* and the various EU negotiators in the Stockholm Convention negotiations declared that they were well aware of how far they could go while still keeping on board all the member states.

5.3.2.2. Causal mechanism As these cases are characterized by a high degree of external compellingness, the agent enjoys a certain degree of discretion, which reveals itself in the exclusive representation role of the EU negotiators. However, enjoying a moderate degree of discretion, the agent is not able to exploit the external compellingness towards a high degree of discretion in a decision-making process in which all information is symmetrically divided between principals and agent and in which the issues at stake are highly politicized. Indeed, the agent has no information benefit vis-à-vis the principals, which means that the EU negotiator cannot employ this as a strategic tool in his relation with the principals in order to extend his discretion. The agent is, furthermore, confronted with principals without private information. This makes it impossible for the EU negotiator to extend the final win-set of the collective principal by transmitting the existing external compellingness because the member states do not have fallback positions. The agent is thus not able to exploit the external compellingness and to increase the political cost of no agreement, as the principals do not have a hidden ultimate win-set. The absence of information asymmetries is, finally, supplemented by a high level of politicization, as a result of which the principals aim to maintain control on the outcome of the international negotiations and, as a consequence, also on their agent.

6. Conclusion

This chapter showed that the compellingness of the external environment is a decisive factor in explaining the occurrence of discretion for the EU negotiator, without being able to judge the particular degree of discretion (Delreux, 2009a). A high degree of external compellingness and the presence of discretion for the agent are closely bound up with each other. In a decision-making process with no (or a low degree of) external compellingness, the EU negotiator's degree of discretion will be low. The opposite goes as well, meaning that, if a decision-making process is characterized by a high degree of external compellingness, the agent will enjoy some discretion, although it is not possible to assess the precise degree of discretion on the basis of the condition 'compellingness of the external environment'. The mechanism behind the link between the external negotiation context and the agent's discretion corresponds to the first extension of the principal–agent model I discussed in Chapter 3. By emphasizing the crucial role of the external context for understanding the internal principal–agent relations in the EU, I join the findings of scholars like Kerremans or Billiet (Kerremans, 1996b; Kerremans, 2003; Billiet, 2006; Billiet, 2009), who argue that the international context has to be taken into account to explain the behaviour of a European agent and the principal–agent relation in EU decision-making processes with regard to the external dimension of first pillar policies.

However, the external compellingness does not explain the whole picture, as it is only able to clarify the presence or absence of discretion. To explain the particular degree of discretion (moderate or high), a more complicated explanation and more qualified causal paths are needed. Decision-making processes with a high degree of discretion for the agent can either be explained by (1) a high degree of institutional density and a low degree of politicization, by (2) a high degree of institutional density and diverging preferences among the member states that have private information, or by (3) a politicized decision-making process in which the agent has information that is not shared with the principals. A final conclusion is that a decision-making process characterized by an agent without private information, combined with either the absence of institutional density or a high degree of politicization and no information benefit for the principals, leads to a moderate degree of discretion for the EU negotiator.

Chapter 6
Unravelling the EU Decision-making Process

1. Introduction

In the previous chapter, the empirical data was analysed in function of explaining the degree of discretion the EU negotiator enjoys in international negotiations leading to an MEA. The purpose of the current chapter is to present general tendencies in the EU decision-making processes regarding such negotiations. While the final section of the previous chapter returned to the empirical data of the separate cases, the current chapter goes back to the variables. By elaborating on three broad themes, I discuss the way the EU decision-making process works in practice. In Section 2, some general conclusions about the way the EU negotiation arrangement is organized are presented. Section 3 focuses on the EU negotiator's discretion. It answers the questions of how the control mechanisms generally function, whether the EU negotiator can conquer more autonomy, and what can be learned from the studied decision-making processes regarding the single voice question. In Section 4, I examine general tendencies in the role the various condition variables play in shaping the EU decision-making process.

2. The EU negotiation arrangement: who is the agent?

As I argued in Chapter 2, from a legal point of view, the EU negotiation arrangement used in negotiations leading to mixed agreements – like MEAs – should be characterized by a clear boundary between the representation of the EC/EU (negotiating the issues falling under EC/EU competences) on the one hand, and the representation of the member states (negotiating the issues covered by national competences) on the other hand. In practice, however, such an arrangement is never consistently followed. The division of competences is only one factor determining the negotiation arrangement of the European Union, besides pragmatism and practicability. Generally speaking, the division of competences is merely a starting point for the organization of the EU negotiation arrangement. On the basis of the legal situation, a practical *ad hoc* negotiation arrangement, which is considered more useful and feasible in the given circumstances, usually develops in a gradual way. Consequently, the question 'who is/are acting as agent(s)?' is most appropriately answered by referring to a combination of legal and practical considerations by the actors in the EU decision-making process.

The division of competences is especially important in determining the role of the Commission. On the one hand, if there is no legislation at the European level on the topics discussed internationally, no negotiation role is granted to the Commission. The negotiations resulting in UNCCD and the Kyoto Protocol, on which EC competences were very limited and where the Presidency and the troika acted as agents, illustrate that the lack of competences prevents the Commission being a EU negotiator. On the other hand, when the issues negotiated at the international level are already (partly) regulated in the EU, the Commission claims – and is granted – negotiation authority. In the negotiations on AEWA, the Aarhus Convention, the Rotterdam PIC Convention, the Cartagena Protocol, the Stockholm POPs Convention and the SEA Protocol, the Commission was granted negotiation authority under article 300 TEC. However, this does not mean that negotiation authority was exclusively granted to the Commission, as the Presidency and/or lead countries could in some cases still act as an agent as well.

The EU negotiation arrangement, as used from BSWG 2 to 6 in the Cartagena Protocol negotiations (i.e. during the largest part of *CART_1*), most appropriately corresponds to the negotiation arrangement that one would expect from a legal perspective. The gentlemen's agreement between the Commission and the Presidency stipulated a division of labour that largely reflected the division of competences. In such a situation, a clear *ad hoc* definition of the division of competences for the issues with which the EU will be confronted at the international level thus is a prerequisite for an EU negotiation arrangement with separate roles for the Commission on the one hand and the member states or the Presidency on the other hand.

The division of competences only determines some basic principles of the EU negotiation arrangement. However, the way the negotiation arrangement is precisely fleshed out depends on practicability and pragmatic considerations. This seems to be particularly the case in decision-making processes regarding negotiations with a more multilateral and even global scope. These dynamics are indeed less prominent in the (regional or intercontinental) negotiations leading to AEWA, the Aarhus Convention and the SEA Protocol. The following examples from the more global negotiations reveal that the EU negotiation arrangement is often *ad hoc*, and not merely explainable by the division of competences, but determined by pragmatic reasoning among the EU decision-makers: the larger negotiating role for the Commission (even beyond its legal competences) in the PIC negotiations; the closely involved member states accompanying and *de facto* dominating the Presidency in the final meetings of the UNCCD negotiations; the evolution from the Presidency to the troika as agent in the negotiations leading to the Kyoto Protocol; the growing role for the Commission in the Cartagena Protocol negotiations; and the lead country approach introduced in the course of the Stockholm Convention negotiations.

One of the most prominent expressions of the pragmatic organization of the EU negotiation arrangement is the deployment of personal capabilities, expertise and know-how from the various actors in the EU. Indeed, pooling the experts of the 12, 15 or nowadays 27 member states is often considered to be an enormous opportunity for the EU to spread tasks among the member states (and often the Commission). This mainly reveals itself in the EU negotiation arrangement used in the contact and working groups, where technical issues are dealt with. In such situations, the EU negotiator is often the actor who is considered the best and most suitable representative for the EU in the given circumstances, regardless of his affiliation. Member state and Commission officials describe this dynamic as follows: 'We have individuals in the EU who are very good in X and others who are very good in Y. We just use them and we deploy them when they are needed,' or: 'We often have persons who are an absolute whiz in a particular matter. It would be stupid if we would send in anybody else. Whether he or she is from the Commission or a member state, it is right that we use this person for that matter.'

The lead country approach is the most noticeable manifestation of this system. It was most visibly used in the negotiations on the Stockholm Convention, but nowadays it is used in other international environmental negotiations as well, with climate change, biodiversity and CSD (Commission on Sustainable Development) negotiations as the most prominent examples. Three reasons explain why the lead country approach and the informal division of labour is used in the EU.

First, the multitude and complexity of the issues at stake in the international negotiations often make it rather difficult for one EU negotiator (Presidency, Commission) to deal with the whole range of issues. When there is a pool of member state experts who are bound by the EU negotiation arrangement and who do not have to represent their member state at the international level themselves, the EU negotiator can appeal to their capabilities, expertise and know-how.

Second, this is a win-win situation for both the agent and the principals. On the one hand, using the assets available at the EU level makes it more feasible for the agent, who is not on his own anymore at the international level. Moreover, by involving the member states in the international process, the agent assures that the principals experience the international negotiation context and the compellingness, which can make it easier for the agent to avoid an involuntary defection. On the other hand, member state representatives feel that they can usefully deploy their assets and capabilities instead of merely being active at the EU level and being constrained by the EU negotiation arrangement to remain inactive at the international level, where the EU negotiator is representing them.

Third, this may increase the EU's bargaining power. If the EU is being dominated at the international level by the external negotiation partners and if the negotiations are going in the wrong direction for the EU, mobilizing all available capabilities and action capacity may help to improve the EU's effectiveness. A Commission representative expressed this reasoning as follows: 'You can be as fanatical about competences as you want, but if this leads to the fact that you are swept away at the international level, you have to be pragmatic.'

The lead country approach only seems to be used in the EU negotiation arrangement when the actors in the EU operate as a group. This means that the member states share the interest in a strongly performing EU during the international negotiations. Two conditions have to be fulfilled for this common interest. On the one hand, the member states need to feel the necessity to form a strong negotiation bloc vis-à-vis the external negotiation partners. The EU position vis-à-vis the Miami Group during the Cartagena Protocol negotiations, vis-à-vis the US in the Stockholm Convention negotiations, or nowadays vis-à-vis the US and the emerging economies like Brazil, Russia, India and China in the current climate change negotiations was/is a catalyst for the EU to present itself as a strong and unified negotiation bloc at the international level. On the other hand, internal factors also need to be present to create a group feeling. First, a dense partnership, a strong team spirit atmosphere and thus a dense institutional environment in which actors trust each other seem to be prerequisites for a group-feeling in the EU and for delegation to lead countries. Second, this can only be the case if the EU decision-making process is characterized by homogeneous preferences among the member states. Only if this is the case can the EU be perceived as a unified group by the negotiation partners.

The way the EU operates in the formal plenary settings mostly differs from the negotiation arrangement used in the technical and more informal contact and working groups. In plenary, the EU position, if any, is mostly represented by the formal EU negotiator, i.e. the Commission and/or the Presidency. If an official from a lead country takes the floor on behalf of the EU, it is usually after the EU negotiator asked the floor and immediately passed it to the lead country representative, who had taken a seat behind the flag of the EU negotiator. At the technical level of the contact groups and working groups, the more informal lead country approach is used more frequently. 'We use expertise rather than flag in these cases', according to a member state official.

3. Discretion, delegation and control in the relation between the principals and the agent

3.1. The principals' perspective: granting autonomy

The degree of autonomy, granted by the member states to the EU negotiator, is determined by the extent and the way the control mechanisms (the contestation of the agent's negotiation authority, the mandate, member states attending the international negotiations, the EU coordination meeting, non-ratification, and repeated delegation) are applied. More in particular, the more the control mechanisms are deployed, the less autonomy is granted to the EU negotiator.

3.1.1. Contestation of negotiation authority
Only in one of the eight studied decision-making processes was the agent's negotiation authority formally contested. From the first to the seventh session of the Aarhus Convention negotiations, the Commission was not authorized to negotiate, although it was clear that EC competences were at stake. However,

the contestation by the member states of the Commission's authority was not merely meant as a control mechanism in order to limit the latter's discretion. The authorization was contested because there was no legal clarity about the impact of the future Convention on the Community institutions.

Nonetheless, the Aarhus case seems more the exception than the rule. Indeed, member states do usually not contest the EU negotiator's authority, neither for the issues touched upon by EC competences nor for the issues covered by national competences. In practice, the non-granting of negotiation authority to an EU negotiator is not considered a realistic option and conducting the negotiations via the EU level mostly goes without saying. Member state representatives declare that granting negotiation authority, even when there are no EC competences at stake, is 'never challenged', 'intrinsic to Europe', 'the practice to do so', or 'never an issue'. Others described this as follows: 'It is clear that it is an EU process', 'it is obvious that we negotiate as the EU', or 'there is no choice, that is how things are, but it is not in the Treaty'. In practice, the authorization *as such* occurs seemingly routinely. Though, as the empirical data indicated, this only relates to the appointment of an EU negotiator, not to the exclusivity of its authority to negotiate at the international level.

Member state officials tend to see the fact that the negotiation process passes via the EU level and via an EU negotiator as self-evident. The reasoning seems to be that 'it is always done like this' (member state representative). Multilateral environmental negotiations usually take place in an institutionalized international setting, in which the EU has already developed a particular *modus operandi*. Some are negotiated at a COP or MOP of their mother treaty (e.g. the Kyoto, Cartagena and SEA Protocols), in which the EU negotiation arrangement was already developed. Others follow a clear UNECE or UN logic. Hence, since negotiations leading to an MEA are not conducted in an institutional vacuum at the international level, the broad outlines of the EU negotiation arrangement are usually adopted from previous negotiations. As a result, conducting the negotiations via the EU is not questioned because it is considered common practice by the member states.

In this sense, the institutional design and the fact that the process goes via the EU level seems to be path dependent (Pierson, 2000; Kay, 2003). The institutional setting in which the decision-making process takes place is determined by previous choices. When these choices are considered good or legitimate (the SI account) or when changing them comes at a high political cost for the member states (the RCI account), they are taken as the basis for the actual institutional design and the authorization. In this sense, when multilateral environmental negotiations are announced, member states consider it a standard operating procedure (Hall, Taylor, 1996; Aspinwall, Schneider, 2001) to participate in these negotiations via the EU level and not in their capacity of sovereign (UN member) state. However, this does not say anything about the degree of autonomy granted to the EU negotiator.

3.1.2. Mandate

Following article 300 TEC – or nowadays article 218 TFEU –, the Commission negotiates the issues falling under EC competences on the basis of a mandate ('negotiation directives') given by the member states. I defined the mandate as an *ex ante* control mechanism in which the member states give procedural and substantive instructions to their agent.

A mandate traditionally consists of two parts: one dealing with procedural aspects, the other with substantive provisions that the member states want to see included in the future MEA. Usually, four procedural instructions are established in the mandate. It stipulates that the EC/EU will participate in the negotiations for matters covered by its competences, that the Commission will negotiate on behalf of the EC/EU in consultation with a committee of representatives of the member states, that the Commission has to report to the member states, and that the Commission, the Presidency and the member states have to cooperate closely on the issues falling under shared competences. This part of the mandate can thus be considered as the formal instrument to authorize the Commission as agent. The procedural aspects of the mandate largely correspond to the provisions of article 300 TEC/218 TFEU. In most of the mandates, even the same language as in the Treaty is used.

After the procedural instructions, provisions about the content of the future agreement are added. There are usually two substantive instructions. On the one hand, appropriate provisions allowing the EC/EU – generally as an REIO – to become a party to the MEA have to be included in the international treaty. On the other hand, the provisions in the MEA may not be incompatible with existing secondary legislation. Hence, the only substantive restriction on the agent's autonomy given in the mandate is defining the existing legislation as the 'red lines', which may not be crossed by provisions in the MEA. Accepting an international agreement that is incompatible with European legislation would imply a renegotiation of this legislation at the EU level. In such a case, a delicate internal compromise would be put at risk and Pandora's box would probably be reopened. The substantive instructions in the mandate are limited to the definition of these red lines. However, inside these borders, the mandate usually grants a large degree of autonomy to the agent, as the EU needs flexibility to engage in the negotiations and to reach an international agreement. Its *de facto* control function is thus rather limited.

If the Presidency or a lead country is authorized as agent, they are not granted a mandate (article 300 TEC/218 TFEU only applies to the Commission as agent), but they also get substantive instructions, which mostly take the form of position papers (see further). However, the main difference in the authorization stage between authorizing the Commission and authorizing the Presidency or a lead country is that the procedural instructions for the Commission are formally written down in the mandate, while this formalized procedural control mechanism is not applied to agents that are a subset of the principals (i.e. the Presidency or a lead country).

3.1.3. Member states attending the international negotiations

In practice, it is a general rule that member states are able to attend the negotiation settings at the international level. In plenary and in the various working and contact groups, the attendance of the principals is taken for granted. Member state representatives can generally freely walk in and out of the international negotiation rooms. In plenary, they mostly sit behind their own national flag; while in contact and working groups, the member states generally take their place behind the EU negotiator. The fact that the member states attend the international negotiations generates a situation of cooperation and control.

On the one hand, such a setting enables on the spot coordination among principals and agent. This on the spot coordination occurs in two directions: member states provide the EU negotiator with ideas, additional arguments or even informal instructions, while the agent can also ask for immediate feedback from the principals. In this way, the presence of the member states at the international level allows employing the expertise and know-how of multiple actors within the EU. If this cooperation occurs in a cooperative spirit and if principals and agent agree on a single European line, this is likely to increase the perceived bargaining power of the EU by the negotiation partners around the table.

On the other hand, principals are constantly able to observe the negotiation behaviour of their agent, which functions as a control mechanism. Member state officials admit that they usually closely follow the statements made by the EU negotiator. Even when these are on paper, member state representatives follow the text. The agent is thus constantly kept an eye on, which reduces his discretion. In this sense, attending the international negotiations is a very effective control mechanism for the principals. An experienced EU negotiator described this control as follows: 'It is horrible negotiating for the EU because it is like having fifteen mothers-in-law sitting right behind you. If you do not say the right thing, there is a very quick tap on your shoulder immediately. It is a very tense situation.'

The so-called Friends of the Chair meetings are the most prominent exception to the general rule that member states can attend the international negotiations. These are limited negotiation settings, convened by the chair of the international negotiations, usually to discuss the most difficult issues. Mostly, the EU is invited to participate as it is usually one of the main players. In such a setting, the EU is represented by the agent, and the member states are not able to attend. However, this claim has to be qualified: the member states are mostly not able to attend these restricted negotiation settings *in their capacity as principal*. Nevertheless, it rarely happens that no officials from a member state are present during a Friends of the Chair meeting, as the (co)chair or the rapporteur of the various contact/working groups, who mostly attend those meetings, often come from an EU member state.

3.1.4. EU coordination meeting

Two kinds of EU coordination meetings take place during the process of negotiating an MEA: in Brussels and on the spot, i.e. during the weeks of the international negotiation sessions. Although there is a large degree of variation among the cases in the frequency of the EU coordination meetings, member state and Commission representatives usually meet in Brussels in-between the negotiation sessions. These EU coordination meetings take place in the setting of a the Council Working Party – more in particular the WPIEI –, which composition, in terms of personnel, varies according to the issue area of the MEA. Recently, these coordination meetings in-between the negotiation sessions are often complemented with consultation by e-mail, in which member states can electronically react to proposals by the EU negotiator.

During the negotiation sessions, coordination meetings are held at least once a day, but often twice or even more.[1] As a general rule, the EU coordinates every day in the morning, mostly one hour or one hour and a half before the international negotiations start. Another coordination meeting is usually organized at lunchtime and if necessary another one in the evening. On the spot coordination meetings are usually organized more frequently and less structured as the negotiation process evolves. The discussions and the (text) proposals at the international level then become increasingly concrete, which requires the EU to react faster and to coordinate immediately.

Besides the general on the spot coordination meetings, experts from interested member states and the EU negotiator often meet in so-called 'breakout coordination meetings'. These meetings are organized to coordinate one set of issues among the officials from the member states and the EU negotiator who have interests and/or expertise in these issues. This means that not every member state is represented in (all) the breakout coordination meetings, although these meetings are formally open for all EU actors. Which member states participate depends on the interests, expertise, capabilities and internal division of work of the national delegations. Such breakout coordination meetings can take place whenever it is considered necessary. As long as the breakout coordination meeting assesses that the EU negotiator on a particular issue can continue negotiating without diverging from the previously agreed EU line, it does not necessarily go back to the general coordination meeting. However, if the officials in the breakout coordination consider a position or a strategy that diverges from or affects the general EU line, the issue is first brought to the general EU coordination before this new position or strategy can be used at the international level. This gives member states that do not participate in the breakout coordination meetings the opportunity to have a say in the EU position on that issue. However, as the member states with a particular preference intensity on the issue usually participate in the breakout coordination meeting on that issue, the position agreed in the breakout coordination is seldom rejected by the general coordination meeting.

In practice, the EU coordination meeting fulfils a double function: determining the EU position and sharing information. Concerning the first function, the cases show that the EU position is determined in an interaction between the agent and

---

1   Only during the first negotiation sessions of the Aarhus Convention were there days without EU coordination meetings.

the principals. From a theoretical perspective, one would expect a bottom-up dynamic, in which the member states provide the EU negotiator with instructions. However, in practice, it is most often the other way round. The EU negotiator suggests an instruction and the member states agree or add something. A decision-maker expressed this as follows: 'I would not say that the member states give an instruction. The EU negotiator has a better feeling of what can be done.' The second part of this quote may seem surprising, as the member states can also attend most of the international negotiation settings. However, two elements in the decision-making process provide the agent a comparative advantage vis-à-vis the member states in assessing 'what can be done'. First, the agent is mostly the contact point for the negotiation partners when they want to assess their proposals or positions with the EU. This mostly occurs informally in the corridors, but also in more formalized outreach activities by the EU. The EU negotiator does not have the monopoly of external (informal) contacts, but it certainly has the most. This provides the agent with the most complete overview of the situation and the different positions at the international level, which is an appropriate starting point for tabling position proposals in the EU coordination meeting. Second, the EU negotiator is usually best acquainted with the progress of the different issues and the developments in the various contact/working groups. As the EU negotiator's negotiation team is usually the largest in the EU, it is able to closely follow the various negotiation settings, while the principals are usually only able to attend one or some of them. This enables the EU negotiator to have the most adequate insights in possible package deals or links between the various issues.

The agent thus mostly determines to a large extent his own instructions, as he generally proposes them to the principals. Indeed, the agent does not only represent the member states at the international level, he also initiates the decision-making processes leading to the EU positions by tabling proposals in the EU coordination meeting. An EU negotiator described this process like this: 'We are presenting the position. Then there are some reactions from the member states and then we adapt the position before going to the international negotiations.' In this sense, the EU position determined in the coordination meeting arises from an interplay between the agent, who assesses and proposes, and the principals who instruct on the basis of these proposals. This does not seem to be typical for environmental negotiations only, as this interactive nature of the determination of the EU position has also been observed in the EU's external trade policy-making (Kerremans, 2004b; Frennhoff Larsén, 2007).

The EU negotiator usually prepares position papers on each topic, in which the historical background, the issues for consideration and the proposed line to take are included. These position papers are constantly updated throughout the negotiation process on the basis of the reactions of the member states and on the basis of the developments at the international level. They function as a starting point for the more concrete instructions the agent uses at the international level. Generally speaking, these instructions can take four forms, going from loose to strict. First, principals and agents can discuss the issues without an outcome on paper. Second, the agent can get a bulleted list with elements that can be used in the negotiations, and from which the EU negotiator can choose the elements he wants to use at the international level. Third, the EU coordination meeting can provide the agent with speaking points or statements for its interventions during the international negotiations. Finally, the EU position can take the form of a proposal for an article that the EU would like to see inserted in the MEA.

The EU coordination meetings also serve as a forum to share information, which is their second function. The on the spot coordination meetings usually start with a debriefing by the EU negotiators about the developments in the various negotiation groups. Moreover, if there are actors from the EU holding an official function at the international level, such as members of the Bureau or (co)chairs of the contact groups, working groups or the plenary, they notify the coordination meeting about the developments at the international level. This is a strategic benefit for the EU, as it can make use of the information coming from the actors from 12/15/27 member states about the state of play of the international negotiations.

In many negotiations, various processes are happening simultaneously, which makes the overall negotiation process too busy for the member states to be able to follow everything. This means the international negotiation process is often non-transparent for a lot of member states. In this framework, the debriefing in the coordination meetings is instrumental for the principals to keep up with the international negotiation process. A member state official acknowledged that 'the coordination meeting is the ideal moment to get to know what happened at the international level'. This does, however, not mean that the flow of information between agent and principals is optimal and that no information asymmetries exist anymore. The coordination meeting only provides a forum to share information, without guaranteeing an equal division of information in the end.

3.1.5. Non-ratification

Not ratifying or threatening not to ratify the MEA at the EU level was never employed as a control mechanism in the cases studied. This corresponds to recent finding in studies of EU decision-making processes with regard to trade negotiations (Dür, 2006). Elsig even refers to non-ratification as 'the nuclear option that is seldom effective' (Elsig, 2007b, p. 12). The ratification decision *as such* is almost always considered as a formality. The statement by a Commission official that 'once you have an agreement, there is in practice no other option than to ratify' was confirmed by a lot of decision-makers, also from the member states.

The control mechanism of non-ratification, namely principals rejecting the commitment that was made by the agent on their behalf, *de facto* moves from the ratification stage to the end of the negotiation stage. In other words: not accepting the political agreement at the end of the last negotiation session is a more realistic and feasible control mechanism than not ratifying. However, in practice, also not agreeing on the final text comes at an enormous political cost for the EU. Everything is done to make sure that all member states and the Commission accept the agreement reached at the international level. If this fails, 'then you really have a problem as negotiator', as a Commission official stated. A solution to such a problem – which did not occur in the decision-making processes studied – would be to convene a Coreper meeting, where the problem could then be discussed at ambassadorial level. As this could take a few days in practice, it would definitely mean a postponement of the conclusion of the international negotiations and a loss of face for the EU vis-à-vis the international negotiation partners.

It mostly takes a relatively long period between the signing of the MEA and its ratification by the EC/EU. In the cases studied, the EC only ratified the MEAs after an average of four years and eleven months. This can be explained by the fact that the Commission mostly initiates a proposal for the ratification decision together with a proposal for the implementation of the MEA. Indeed, the Commission's general reasoning seems to be that an MEA can only be ratified when it is clear how it will carry out its international commitments made in that MEA. A Commission official phrased this as follows: 'You cannot ratify something if you do not know how to implement it.' The preparation of a legislative proposal, including consultations with expert committees and stakeholders, always takes some time, which also explains the relatively late ratification.

The ratification decision is thus initiated by the Commission. In this proposal, the Commission usually briefly describes the content of the MEA and some factual observations about the negotiation process (number of negotiation sessions, the external negotiation partners involved, etc.), and it proposes the text for a Council decision on ratification. As the consultation procedure was applicable in the period studied in this book, the EP could advise the Council on the ratification.[2] Also the Committee of the Regions and the European Economic and Social Committee can advise on the EC ratification. Among the cases studied, they only did so for the Stockholm Convention. In all cases, by contrast, there was an advice by the EP, mostly consisting of a description of the background of the MEA and the negotiation process, the Parliament's evaluation of the agreement, and a recommendation on the ratification. For a couple of the ratification decisions, the EP proposed some amendments to the legal text. Finally, the Council decides on the ratification. Such a ratification decision consists of four parts: it lays down the legal basis of the decision, it includes a number of general considerations, it has an article stating that the EC approves the agreement and another authorizing the Presidency to designate a person to deposit the ratification instrument. Some ratification decisions have a number of additional articles, mostly dealing with a particular provision of the agreement. Finally, the complete text of the MEA is incorporated as an annex to the decision and published in the Official Journal.

Apart from fulfilling its international commitments, the EC/EU has another incentive to ratify: ratification is needed to be able to participate as a full party, implying full voting rights, to the subsequent COPs or MOPs of the MEA.

---

2   As discussed in Chapter 2, the role of the EP has been considerably enforced since the entry into force of the Lisbon Treaty. Nowadays, the EP has to give its consent – and no longer its advice – on an MEA before it can be ratified.

Likewise, the individual member states are urged by the Commission to ratify the MEA at their domestic level. Only the parties to the MEA, i.e. the countries that have ratified, have a formal vote in the COPs and MOPs. If the EU then manages to vote in a coordinated way, its voting power is increased with each member state already being a party.

3.1.6. Repeated delegation

The agent's discretion and principal–agent relations do not seem to have a control spill-over effect from one decision-making process to another. Only in the PIC negotiations does the control mechanism of repeated delegation seem to have played a role. It was not explicitly employed by the member states (they did not grant less autonomy to the EU negotiator in a future negotiation round because of the negotiation behaviour of the Commission during the PIC negotiations), but Commission officials took into account that their negotiation behaviour could possibly influence the autonomy that member states would grant them in future negotiations. In none of the other decision-making processes studied did any decision-maker recognize this control mechanism.

3.1.7. Controlling the agent?

The evaluation of the six control mechanisms shows that the *ad locum* control mechanisms, i.e. the coordination meeting and the member states attending the international negotiations, play a more important role in determining the relation between principals and agents than the *ex ante* and *ex post* control mechanisms do (Delreux, 2008; Delreux, 2009d). The *ad locum* control mechanisms are the most effective control mechanisms, a finding that also emerges from EU trade policy-making analyses (De Bièvre, Dür, 2005; Elsig, 2007b; Elgström, Frennhoff Larsén, 2010). As the authorization is usually not contested, the mandate broad, non-ratification not used to call back the EU negotiator, and the future negotiation rounds not taken into consideration, the control function of these mechanisms was much more limited.

It is of course still possible that the agent's negotiation behaviour and his relation vis-à-vis the principals are determined by the anticipation on the ratification stage. In other words: the observational equivalence problem may pop up here (see Section 4.1 of Chapter 3). If this is the case, the ratification requirement would be the most prominent control mechanism. Although it is difficult to put forward a completely sound argument and to come up with hard and absolute evidence to prove that the agent's anticipation on the control mechanisms in the ratification stage was not the decisive factor in limiting his discretion, I assess this as very unlikely on the basis of my (interpretation of the) data. The interviewed principals and agents always referred to the *ad locum* control mechanisms if they talked about the discretion of the EU negotiator. EU negotiators seemed to be mainly concerned to get the principals involved in the international negotiation process and to get them on board during the international negotiations. They needed the backing of the member states on the spot and during the coordination meetings.

Moreover, the agents seemed to assume that they would not be confronted with surprises in the ratification stage.

The two *ad locum* control mechanisms do not only fulfil a control purpose. In fact, they function as a dynamic interplay of cooperation and control (Murphy, 2000; Kerremans, 2004b; Delreux, 2009d). The fact that member states attend the negotiations at the international level certainly limits the EU negotiator's discretion. It thus functions as a control mechanism from a principal–agent point of view. But at the same time, the on the spot coordination also increases the cooperation between principals and agents. The same observation can be made with regard to the coordination meeting: it is an effective control mechanism because it urges the agent to frequently give account to the principals and because it allows the principals to instruct their agent, but is also a forum for cooperation between member states and EU negotiator, since the principals and the agent are usually looking for an appropriate solution for the EU and since instructing (and thus controlling) the agent is not a one-way process.

The agent usually negotiates on the basis of instructions, which are gradually developed during the decision-making process. This allows the principals and the agent to anticipate on the international context in which the EU needs to (re)act. The initial stages of the decision-making process are usually characterized by a large extent of uncertainty about how the international negotiations will evolve and about the proposals that will be tabled at the international level. Therefore, it is rather impossible for the member states to provide their agent with tight instructions.

Generally, three steps can be distinguished in giving instructions to the agent. First, the mandate for the Commission determines the red lines, which cannot be crossed and which usually correspond to existing European legislation. Second, in some negotiations, Council conclusions formally refine the mandate. As Council conclusions can be issued repeatedly, they can precis the mandate because they are taken at ministerial level as well. Regularly issuing Council conclusions was used to update the general instructions in the EU decision-making process with regard to the Kyoto Protocol, Cartagena Protocol and Stockholm Convention negotiations. Third, when the negotiations proceed, this general framework of the mandate and the Council conclusions is usually readjusted and specified in position papers, speaking points or even proposed treaty texts. A Commission negotiator described this process as follows: 'The line of the Commission is always to have a broad mandate and then to refine it through further discussions with the member states throughout the negotiation process'.

3.2. The agent's perspective: conquering autonomy?

An agent conquering more autonomy against the wishes of the principals – or 'agency slack', in theoretical terms – seems to be more the exception than the rule. Only in the last days of the Kyoto negotiations did the agent conquer additional autonomy. However, this does not mean that an agent never *enjoys* more discretion than the autonomy granted by the principals. At the end of the negotiations, the EU negotiator is often urged by the international context to react quickly, i.e. without prior consultation with the principals. However, such autonomous behaviour by the agent does not come at a cost for the principals, as the EU negotiator does not conquer autonomy at the expense of the member states' preferences. In other words, this cannot be considered as slack.

Member state representatives even admit that they expect that the EU negotiator acts autonomously in such a situation. They want their agent to take his responsibility and to negotiate strongly. Member states expect from the EU negotiator the ability to react on the spot on proposals by the negotiation partners or by the chair of the international negotiations, without rigorously sticking to the speaking points, but still remaining within the general EU line as decided in the coordination meeting. In other words, the agent enjoys some additional autonomy, although not explicitly granted, at the end of the negotiation stage to proceed reactively.

If the agent reacts on (and even accepts) proposals without consultations with the principals, this only occurs at the very end of the process. During most negotiation sessions, it is common practice that the EU negotiator does not accept a particular proposal before having the backing of the member states. If there is no EU position yet on a (new) proposal, the agent usually expresses a reservation at the international level, asking the chair to postpone the issue and to resume it after the next EU coordination meeting. This system of continuous EU coordination mostly allows the EU negotiator to get all the member states on board and to be able to present a common position, which increases his bargaining power. However, besides this benefit, this mode of operation has two disadvantages. On the one hand, it reduces the flexibility of the EU, which is not only often an impediment in international negotiations, but which is also 'quite annoying for the negotiation partners', according to an EU negotiator. On the other hand, continuously coordinating is very time-consuming. The time spent in coordination meetings cannot be used for contacts with the negotiation partners. Indeed, before the international sessions start, at lunchtime and in the evening, the negotiation partners are never inactive. When they are informally discussing the issues in the corridors, the EU is not involved in these discussions and, consequently, it cannot participate in the (package) deals made there. As an EU negotiator phrased it: 'The more you coordinate, the less you negotiate'. This is not only the case in international environmental negotiations, as it is also one of the conclusions in studies on the EU's role in international negotiations on trade (Baldwin, 2006; Elgström, 2007) or human rights (Smith, 2006) or the EU coordination at the UN (Degrand-Guillaud, 2009).

3.3. What about the single voice?

So far, I have touched upon the questions of the EU negotiation arrangement and the discretion of the EU negotiator. Knowing who negotiates with which degree of discretion leads to the question whether the EU manages to speak with a single voice or not. The answer is not clear-cut. Generally speaking, there are four ways the EU can be represented, determined by two factors: is there a general EU position or a general EU line that is followed (single voice or multiple voices) and, if so, is this position expressed by one or by multiple actors (single mouth or multiple mouths)?[3] The four possibilities, ranked from unison to division, are presented in Figure 6.1, together with the corresponding cases.

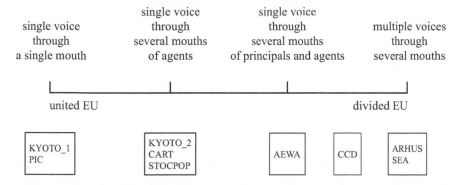

Figure 6.1   Single and multiple voices and mouths in the EU representation

Speaking with a single voice through a single mouth guarantees consistency in the EU position. It excludes that the EU position is perceived as being characterized by internal ambiguities, which could possibly lead to negotiation partners playing the EU voices against each other. Moreover, the member states can prefer the 'single voice through a single mouth' option because this allows them to play the blame-shifting strategy when they are back in their capital. It gives them the opportunity to shift the blame for a particular outcome to the EU level, as they can refer to the fact that they were bound by the common position, which could only by expressed by the single negotiator. However, if only one agent takes the floor – and thus if a single mouth is used –, EU decision-makers admit that it is not unlikely that the negotiation partners will not be sufficiently aware that there are 12/15/27 member states behind that mouth. In other words, by expressing the common position of multiple countries only by one mouth, the intended increased bargaining power can disappear, as it is possibly not clear that this single mouth represents a common position, backed by all member states.

To undo the disadvantage of the single mouth, but to retain the single voice benefit, the 'one voice through several mouths' (Cœuré, Pisani-Ferry, 2003) strategy is often considered the optimal option. There are two possibilities to realize these several mouths: only agents taking the floor, and both agents and principals

---

3   I owe the concepts 'single voice' and 'single mouth' to former trade Commissioner Pascal Lamy, who used it e.g. in his 2002 speech at the Humbolt University (Lamy, 2002).

taking the floor but always defending the same European line. The EU negotiation arrangement can consist of multiple agents (e.g. the troika, or the Commission and the Presidency), which are the only EU actors who can take the floor. If the EU manages to express a common position by a clearly distinguishable team of agents, 'we are almost unbeatable', according to an EU negotiator. Therefore, two conditions need to be present: a clear common position and clearly defined agents. To emphasize that there are several member states – being future parties to the MEA – behind the single EU voice, several mouths are thus considered instrumental.

This system can be extended by allowing the principals to express the EU position as well. If this strategy is used, the agent usually firstly takes the floor to present the EU position. Subsequently, this position is reinforced by the principals, e.g. by giving additional arguments or examples. This strategy also increases the (perceived) bargaining power of the EU, as the rationale is that the point seems to be stronger if it is repeatedly expressed.

A final possibility is to express several (different) positions by multiple mouths. Although the negotiations where this happened (on the Aarhus Convention and the SEA Protocol), were characterized by diverging preferences in the EU, not expressing a single voice can be a considerable option in a negotiation setting where the EU has a large relative bargaining power. In such settings, although it is usually not a clear strategy, it can be beneficial for the EU not to dominate the international negotiations (and thus not to speak with a single voice or a single mouth). At first sight, this may seem counterintuitive, as one would expect that negotiation power and effectiveness (i.e. the extent to which the EU position is reflected in the outcome of the international negotiations) are positively correlated. However, negotiating via a single voice or a single mouth may turn out counterproductive if the relative bargaining power of the external negotiation partners is limited. The few negotiation partners may have a sense of having no say in the negotiations. It takes two to tango – and in a multilateral, although regional, setting even more – and if the EU wants to reach an agreement at the end of the negotiation process, it has to keep its negotiation partners on board. A single voice or a single mouth, which would imply a smaller role for the member states and a perceived strong and united EU, is neither desirable nor instrumental in such a situation.

4. The conditions of the EU decision-making process

4.1. Preferences

4.1.1. Preferences of the principals
The theoretical situation of each member state having a clear preference, derived from a national interest and expressed at the EU level, does not always completely correspond to reality. This can be explained by two elements: not all players in the

collective principal seem to have – or at least to express – a preference, and if they have one, it is not always a genuine national position.

First, not all member states are equally actively involved in the EU decision-making process. A lot of them do usually not – or only very seldom – take the floor in EU coordination meetings. Mostly, there are only a few member states that are closely involved and express their preferences. Others do not (or do only very occasionally) take the floor in the coordination meetings. They seem to agree with what has been said and to acquiesce in the outcome of the meeting. The intensity of the member states' preferences thus seems to vary. Although there is a large degree of variation among the decision-making processes on which member states were active and which member states were less involved,[4] member states such as Germany, Denmark, the Netherlands, Sweden or the UK were mostly – if not always – closely involved and active at the EU level.

Second, it may be the case that the preferences expressed by a member state official at the EU level largely correspond to the personal opinion of this particular official. When technical issues are discussed in decision-making processes, representatives of certain member states seem to express positions more in a personal capacity or in their capacity as an expert than as a representative of (the preference of) their member state. A Commission official expressed this observation as follows: 'Do not have illusions. On technical matters, the position of a member state is often the opinion of the key person who is there during the international negotiations.' Moreover, if member state officials formally negotiate at the EU level on the basis of national instructions, those were often written by those same officials and only rubberstamped by the Foreign Affairs Ministry. In some decision-making processes, officials of certain member states even acted without a substantive national instruction from their capital, leading to a situation where officials express their personal opinions, officially on behalf of a member state. However, these preferences are considered as national preferences when one looks at it from the perspective of the EU, although they are largely connected to the individual representing that member state.

4.1.2. Preferences of the agents

Except in the EU decision-making process with regard to the Aarhus Convention negotiations, a substantive Commission preference about a particular provision of the future MEA could hardly be found. The only clear preference of the Commission is that the MEA needs to be compatible with existing secondary legislation. This usually does not diverge from the member states' preferences, as consistency with existing legislation is one of the few elements in the mandate.

---

4   In the comparative overview of the cases in Chapter 4 (see Table 4.2), and more in particular the row 'overview positions EU level' of the table, the most active and the most closely involved member states are portrayed on the preference axis for each decision-making process.

Like the Commission, a negotiating Presidency usually does not defend its own preferences. Generally speaking, when a member state holds the Presidency, it tends to put aside its national preferences. Member state representatives that have been in the Presidency seat confirm that 'when you have the Presidency, you have other responsibilities. You are in the chair and you cannot keep extreme positions'. They stress that their behaviour is constantly determined by the awareness that they are supposed to be neutral and that their task is to focus on the outcome of the decision-making process and to find a compromise between the member states. This finding is in line with the result of various studies about the role of the Presidency, stating that neutrality and putting aside national preferences for six months are inherently connected to the Presidency's role (Svensson, 2000; Tallberg, 2003; Schout, Vanhoonacker, 2006; Schalk, Torenvlied, Weesie, Stokman, 2007; Fernandez, 2008).

4.2. Politicization

The studied negotiations generated a considerable degree of variation in the level of politicization they caused in the member states. While the negotiations leading to UNCCD, AEWA and the PIC Convention were politically completely insensitive in the EU, the Kyoto negotiations and the second half of the Cartagena Protocol negotiations were certainly characterized by a high level of politicization.

The most noticeable manifestation of a high level of politicization is the participation of ministers in the decision-making process. Apart from various signature ceremonies, this only happened in the last sessions of the Kyoto and Cartagena Protocol negotiations. The reasons why ministerial segments are instrumental in the final stage of politically sensitive negotiations, is that ministers have the authority to deviate from Council conclusions, which are often general instructions for the agent. Hence, ministers are in a position to take the ultimate decisions. If the member states were only to be represented at expert, civil servant or ambassadorial level, the EU coordination meeting would not be able to accept an MEA that diverges from instructions issued at ministerial level. Moreover, the attendance of ministers is instrumental in making political deals 'behind the scenes' and in putting political pressure on the external negotiation partners.

4.3. Information

4.3.1. Asymmetrically divided information in favour of the agent
Generally speaking, EU member states are well informed about the substance and the developments of the international negotiations. This is mainly caused by the fact that most negotiation settings can be attended by the principals. Only about the developments in Friends of the Chair meetings, where the chair of the international negotiations or a contact or working group decides who can participate, are member states not always able to rely on first-hand information.

In such a situation where the agent enjoys an information benefit, the tolerance by the member states with respect to the information asymmetry matters. When member states assess that the international process is so busy or so unmanageable for the agent that comprehensive debriefings cannot be provided, it is likely that they will tolerate a certain degree of information asymmetry. This happened in the last stages of many negotiations, the Kyoto and Cartagena Protocol negotiations (*in casu* the Group of Ten) being the most prominent examples. By contrast, when member states get the impression that the EU negotiator intentionally withholds information, tolerance will most likely decrease. However, member state officials and EU negotiators admit that even in such situations the agent seldom possesses private information. They emphasize that sharing information with the member states is in the interest of the agent as well. On the one hand, transmitting information to the member states helps the EU negotiator in creating trust among the principals. If member state representatives notice that the agent intentionally withholds information, the employment of the control mechanisms would immediately be tightened, decreasing the degree of granted autonomy.[5] Because the agent wants to avoid being blamed for suppressing information, the information is usually shared with the principals. On the other hand, EU negotiators state that they provide a good flow of information to the member states because they feel the need for feedback and support from the member states. An EU negotiator expressed this as follows: 'You feel lonely at time, when you are engaged in this kind of negotiations. You need some backing, you need the feeling that you have people behind you who support you. This cannot happen if you do not inform them.'

In assessing the information asymmetry in favour of the agent, it is important to distinguish two separate questions: first, were the member states able to be fully informed?; and second, were all member states fully informed? The answer on the first question is mostly affirmative. Nearly always, the mechanisms to be informed, such as debriefings in coordination meetings or the possibility to attend the negotiations, were present. The second question, on the contrary, needs to be answered negatively. Although all member states mostly have the opportunity to be fully informed, not every member state is. Consequently, whether a member state has the same degree of information as the EU negotiator does not depend on the agent, but on the principal's involvement and on the extent to which the principal makes use of the tools that are at his disposal.

---

5   It is very likely that the member states would notice it if the EU negotiator withheld information. Member states have also direct contacts with negotiation partners during the negotiation process. Prominent examples, noticed in a lot of cases studied, are France having intense informal bilateral contacts with Western African countries, or the UK with Commonwealth countries.

4.3.2. Asymmetrically divided information in favour of the principals

A national preference, if any, seldom remains fixed during the decision-making process, as it is usually constantly adjusted to developments at the international level. In almost all negotiations studied, new ideas and proposals popped up during the international negotiation process. As a result, preferences needed to be developed regarding these new ideas and proposals. In many cases, member states had no clear preferences in the beginning of the decision-making process. They did not have a clear picture on what the MEA would look like. If a member state has no preference, it neither has a fallback position, and thus private information. A member state official expressed this as follows: 'Many member states do not know their bottom-lines at the beginning of the negotiations'.

On the basis of the decision-making processes studied, I can conclude that principals do not always have private information about their fallback positions, but they sometimes do. Question is now: when and under which conditions? The answer seems to be twofold.

First, the position of the member state's preference in relation to the preferences of the other member states seems to determine whether using fallback positions is useful or not. On the one hand, if a member state holds a preference that is an outlier in the EU, it can opt to keep the information about its fallback position private vis-à-vis the EU negotiator. This happened both with member states having minimalist outlier preferences and with member states having maximalist outlier preferences. An example of the former is the French preference on hunting during the AEWA negotiations. The latter is illustrated by the preferences of the Toronto group of member states (Austria, Sweden, Denmark and Germany) on the emission reduction targets in the negotiations on the Kyoto Protocol. In the first situation, the EU negotiator had to assess how much he could *minimally* concede at the international level to keep France on board. In the second situation, the agent was confronted with the question how much he could *maximally* give in. On the other hand, the closer the preference of an individual principal is positioned to the median preference in the collective principal it belongs to, the less useful private information about the fallback positions seems to be. In other words: it does not appear useful for a member state with a non-outlier preference to keep the information about its fallback position asymmetrically divided in its favour.

Second, the degree of asymmetrical information in favour of the principals seems to be linked with the degree of politicization. However, this link is indirect, as it is affected by the degree of preference homogeneity between principals and agent. Indeed, if the decision-making process is politicized and if the preferences of the EU negotiator *correspond* to the member states' preferences, the decision-making process at the EU level is likely to be open and transparent, i.e. not characterized by an information asymmetry in favour of the principals. The second stage of the Cartagena Protocol negotiations is an example of this. Indeed, if a lot is a stake for the principals, but they expect that the same is at stake for the agent, they will not withhold their fallback positions. However, if the decision-making process is politicized and if the preferences of the EU negotiator *do not correspond* to the member states', the EU decision-making process is likely to be characterized by private information for the principals. The Kyoto Protocol negotiations illustrate this dynamic: the troika, with the Netherlands and the UK having somewhat more maximalist preferences, was not fully informed about the fallback positions of the member states. The high level of politicization of the climate change issue did certainly contribute to this.

4.4. External compellingness

One of the results of the QCA procedure (Chapter 5) teaches us that the external compellingness determines whether an agent will enjoy a particular degree of discretion or not. How does this external compellingness work in practice? First, member states certainly experience this compellingness. They describe it as follows: 'You feel the pressure that there has to be a result', 'nobody wants to be the one who makes it fail', or 'there is a hidden pressure to get an agreement'. A member state official even formulated it in more theoretical terms: 'The political costs are too high to use the formal veto power you have'. Moreover, the external compellingness leads to member states revealing their less intense preferences instead of their ideal points as the decision-making process evolves. This is illustrated by the following from a member state representative: 'We have to explain back home that the logic of the negotiations meant that we had to depart from our preferred national position'.

Second, the compellingness can be used as a facilitatory factor for the EU negotiator to conquer more autonomy. European decision-makers participating in the final stage of the Kyoto Protocol negotiations, the only case studied where the agent conquered more autonomy, acknowledge that the fact that 'the member states wanted an agreement was definitely one of the reasons why the EU negotiator could go beyond his mandate at the end'. A troika representative phrased the agent's reasoning as follows: 'I knew that the situation where member states would not agree was a theoretical possibility, but I also knew that it was *de facto* not possible'.

Besides the degree of external compellingness, resulting from the number of external negotiation partners, their relative bargaining power and the way this determines the international negotiation process, EU decision-makers distinguish three other types of pressure on the member states to make the negotiations succeed. First, as negotiations cost money, member states often experience a financial pressure. Not only does each member state have to finance the travel and accommodation expenses of its representatives, but the financial means that are put aside for the negotiations by the international organization under which auspices the negotiations are held (UNEP, FAO, UNECE, etc.) are not inexhaustible. Second, international negotiations, meant to result in an MEA, are often established by a mandate, adopted by the parties of an international organization of the mother treaty.[6] Such a mandate often includes a deadline by when the negotiations should be finished, which creates a time pressure for the negotiation parties, including the EU member states. Third, most MEAs are signed during a meeting at ministerial level following the negotiations.[7] The date for these ministerial segments is usually already scheduled before the end of the negotiations. The fact that there has to be an agreement to present to the ministers often creates a political pressure. As a result, one of the most valuable, yet implicit, functions of a signatory ceremony at ministerial level is to bring the negotiations to a successful end.

4.5. Institutional norms

4.5.1. Institutional memory and diffuse reciprocity

The institutional norm of diffuse reciprocity means that actors base their negotiation behaviour on the understanding that they have to continue cooperating with each other in the future. The negotiation of an MEA requires multiple negotiation sessions. In each session, decision-makers from the member states and the Commission are aware that they are *de facto* obliged to tackle the international negotiations together. Generally speaking, diffuse reciprocity seems to play an implicit, inherent role among the member states, but views differ as to its effect on the relation between the member states and the EU negotiator.

Member states respect the fact that they have to be able to cooperate with other member states in the future. This not only refers to the negotiation process, but also beyond it, e.g. to the implementation of the MEA or to decision-making on internal legislation in the same policy area. Member state officials admit that they are 'in a long term relationship, which is taken into account'. The reasoning seems to be that the joint awareness of their common future leads to searching for joint solutions at the EU level. A member state representative described the functioning of this norm as follows: 'You are doomed to each other. That is a tacit assumption, but it plays a role. If you have a difficult point and if you would have been obstructive in the past, they wipe the floor with you and you are out of the game.'

While diffuse reciprocity thus seems to be inherent in the relations among principals, it is not that clear whether this norm also influences the relation between principals and the agent, in particular between the member states and the negotiating Presidency. On the one hand, some decision-makers confirm that it plays a role, in the sense that member states are aware that they will hold the Presidency in the future. Therefore, they do mostly not impede the negotiating

---

6   Examples are the Berlin Mandate for the Kyoto Protocol negotiations (adopted by COP 1 of UNFCCC), the Jakarta Mandate for the Cartagena Protocol negotiations (adopted by COP 2 of CBD) or the Sofia Declaration for the SEA Protocol negotiations (adopted by the third UNECE Environment for Europe Conference).

7   Of the cases studied, only AEWA and the Kyoto Protocol were signed during the last negotiation session, and thus not in a separate signing ceremony.

Presidency. On the other hand, others state that Presidencies do not take into account diffuse reciprocity. A member state official even phrased it like this: 'A Presidency wants a success and a result at the end of its six-months term and it reasons: "who cares about what member states think about me afterwards?"'.

4.5.2. Trust and credibility

Trust, established in the institutional framework of the EU, seems to play an important role in the relation among the decision-makers at the EU level, in particular between member states and the EU negotiator. Most decision-makers attribute trust to personal relations. They emphasize the importance of knowing the people around the table, who cooperate for a relatively long time. Many key players in the EU decision-making processes knew each other from other international negotiations or from EU decision-making processes on internal legislation in the same policy area. This is said to establish personal trust. Some decision-makers even describe these personal relations in terms of 'friendship' or 'one big European family'. A minority of the interviewees attach more importance to trust in institutions than trust in people: they state that they 'trust the EU negotiator because it is his institutional role to be the representatives of the member states'. This is not to say that trust in the agent, be it in the person or in the institution, is generally absolute. As the international setting of negotiations leading to an MEA mostly allows the member states to take the floor internationally, the principals still maintain a big stick. Except in the Friend of the Chair settings, they have the possibility to intervene and to immediately adjust a possible bad situation when an agent would harm their trust.

Rather logically, the degree of trust in a particular agent depends on the agent's skills, assets and capabilities. The personal resources of the people negotiating on behalf of the EU, including their negotiation skills, fluency in English and substantive knowledge of the issues at stake, largely determine how much they will be trusted by the member states. A precondition for an agent to be trusted is that, according to a member state official, 'member states can feel that they are in good hands'. In this respect, the practice of certain member states to incorporate new officials, who are higher on the national hierarchy but who are also often new in the process or inexperienced in the policy field at stake (ambassadors, etc.), in the Presidency team for the course of their six-month Presidency is not beneficial in creating trust among the principals.

4.5.3. Mutual responsiveness

Mutual responsiveness, i.e. the institutional norm leading to decision-makers knowing and understanding (the background of) each other's preferences, which allows them to frame the issues at stake from the perspective of their colleagues, does not seem to play a decisive role in the studied EU decision-making processes. From the five institutional norms that I distinguished, member state and Commission officials evaluate mutual responsiveness as the least important norm in determining the decision-making process. Expressing justifications and reasonings behind certain preferences has a rather limited effect on other member states, as it only seems to play a role as long as it does not interfere with the preferences of these other member states. Almost all member state representatives acknowledge that their own preference prevails, but that they are willing to go along with reasonings of other member states where it does not touch upon their own preferences. In other words, mutual responsiveness only plays a role in accommodating other member states on issues that do not affect one's own preferences. If it would imply a concession on its own preference, this norm does normally not affect the behaviour of a member state at the EU level. Briefly, there is mostly a certain degree of understanding of each other's positions and problems, but this does not lead to preference changes.

4.5.4. Consensus and compromise striving

Striving for a consensus and a compromise at the EU level seems to be the most prominent institutional norm. Consensus in the EU does not mean that every member state has to be a proponent of a particular proposal. It means that no member state rejects it. A large majority of the interviewed officials assessed the EU decision-making processes as characterized by a high degree of consensus striving. Most of them even see it as something obvious. To substantiate the normative power of consensus and compromise striving, two types of evidence pop up. First, member state officials admit that isolated preferences, mostly on minor issues, can seldom be maintained at the EU level. They phrase this situation as follows: 'If you feel isolated, you mostly drop your resistance, because you are a member of the [c/C]ommunity' or 'it is better not to be the loser'. Second, although some of the studied cases were characterized by heterogeneous preferences among the member states, which impeded the EU decision-making process and the development of a common position, a vote never took place. Most officials admit that voting would have a very bad effect on the atmosphere in the EU. One representative, who experienced a vote in an EU decision-making process during another international negotiation, described the atmosphere of voting in the EU as follows: 'It is terrible, it is really terrible. You have to expose yourself and you get branded. It completely destroys the atmosphere in the EU group.' Hence, voting is *de facto* out of the question, or at least a very last resort. Instead, one relies on so-called *tours de table*: each member state is quasi-forced to put forward its position on a particular issue, but it retains the possibility of qualifying its position, instead of being obliged to limit that position to a yes or a no. If a *tour de table* has taken place in the EU, it is rapidly known by the negotiation partners. On the one hand, it makes clear that the issue at stake is important for the EU. But on the other hand, it also makes the EU vulnerable if it does not manage to gain this point at the international level.

4.5.5. Avoiding process failure

My empirical data showed that the member states certainly aim to avoid a failure of the international negotiation process and the EU decision-making process.

However, this does not seem to be directly caused by the institutional context of the EU decision-making process. The fact that a breakdown of the EU process is not considered an option can be explained by two factors. First, as mentioned above, participating in international environmental negotiations via the EU level is seen as self-evident by the member states. Negotiating separately and passing over the EU level or jeopardizing the EU process is *de facto* not a realistic option. Second, member states usually do not want the international negotiations to fail. As the international and the EU level are closely interconnected, avoiding process failure at the EU level cannot be unlinked from avoiding process failure at the international level. As a result, avoiding process failure at the EU level does not seem to be a goal for the decision-makers, but a means for avoiding process failure at the international level.

4.5.6. The role of sociological institutionalist dynamics in the EU decision-making process

In Chapter 3, insights from SI were theoretically applied to decision-making processes characterized by delegation of negotiation authority and granting of autonomy. This section explores how institutional norms affect the three stages (authorization, negotiation and ratification) of the EU decision-making process with regard to international negotiations leading to MEAs. In this sense, this section verifies and checks the third extension of my principal–agent model, as developed in Section 3.3.3 of Chapter 3, and more in particular Figure 3.8.

In the authorization stage, SI seems to offer explanatory value. Evidence of principals calculating the costs and benefits of the decision whether they will delegate – and, if so, to whom – could not be found. In the dense institutional context of the EU, there is no delegation decision *as such*. Negotiating via the EU level and via an EU negotiator, also for the issues covered by national competences, is neither questioned nor contested. A particular institutional design is applied, without (explicitly) (re)considering it and without making a new cost-benefit analysis on it, because it is considered as a conventional practice by the principals. In that sense, the SI perspective on the decision to delegate seems to prevail. This corresponds to Gilardi's proposition that the institutional design of delegation can be explained by the fact that this particular design is highly valued by the decision-makers participating in the EU institutional context (Gilardi, 2002). As the institutional structure from previous decision-making processes with regard to international environmental negotiations are copied, it can be argued that the institutional environment generates isomorphic pressure on the principals (McNamara, 2002).

In the negotiation stage, principals and agents are well aware of the fact that they find themselves in an interdependent relation. On the one hand, the EU negotiator needs the backing of the member states in order to be able to represent a high degree of bargaining power at the international level. Moreover, the agent needs the approval by the member states of the agreement reached at the international level in order to avoid an involuntary defection. On the other hand, the member states need the EU negotiator to play an important role at the international level in order to have an impact on the outcome of the negotiations. In this interdependent relation, principals and agents seem to base their behaviour on a combination of rational anticipation and institutional norms like trust or consensus striving. In any case, the relation between principals and agents in the negotiation stage is not as instrumentally driven as traditional (RCI) principal–agent theory assumes. Most EU decision-making processes were characterized by a partnership atmosphere between the representing and the represented actors. The extent to which such a partnership atmosphere emerges seems to depend on the degree of preference homogeneity among the member states. The non-occurrence of a partnership atmosphere usually coincides with diverging preferences among the member states. Hence, the lack of a cooperative relation between the principals and their agent can mostly be attributed to a lack of cooperative spirit among the principals. The empirical data thus confirms that the traditional RCI principal–agent framework, based on interest-maximizing actors engaging in a principal–agent relation characterized by control and the possibility of agency slack, needs to be supplemented with insights from SI, which emphasize the power of institutional norms that make the decision-making processes more cooperative.

In the ratification stage, the relations between the principals and the agents and the institutional context (i.e. the SI account) do not seem to be decisive in explaining the behaviour of the member states. However, neither a cost-benefit analysis nor an assessment of the negotiation behaviour of the EU negotiator in the preceding stage seems to determine the ratification behaviour of the member states in the Council. 'Once we have signed, we ratify,' as expressed by a member state official, appears to be the general rule. In that sense, from the perspective of the member states, a signature implies a commitment vis-à-vis the other (external) signatories to fulfil their international responsibilities.

5. Conclusion

This chapter has shown that in a principal–agent situation, it is not only instrumental for the principals to control the agent; it is equally instrumental to engage in cooperation with the agent. While the benefits of control relate to the ability to avoid opportunistic behaviour by the EU negotiator, the benefits of cooperation are that the principals can be involved in the negotiations and contribute to the development of the international agreement, without undermining the functional benefits of delegation. Indeed, the EU decision-making process with regard to international environmental negotiations is essentially characterized by a principal–agent relation because member states opt to delegate negotiation authority to an EU negotiator since they expect functional benefits from such an arrangement, namely an increased bargaining power for the EU. This benefit does not disappear when principals and agents cooperate, and it increases the possibility for the principals to remain an active player in the task that is delegated to the agent.

Moreover, these cooperative dynamics in the principal–agent relation are not only instrumental for the principals, but for the agent as well. If the EU manages to pool all the capabilities (mainly expertise, know-how and information) available in the EU, irrespective of the affiliation of the official who provides these capabilities, it enormously strengthens the EU as an international actor and negotiator. Cooperation between the member states and the EU negotiator strengthens the EU as a negotiation party vis-à-vis its negotiation partners around the international negotiation table. It also acts as a signal to these negotiation partners that there is no internal disagreement in the EU and that the EU operates as a unified bloc. All this means that the agent, cooperatively backed and supported by the principals, is able to negotiate with a high degree of bargaining power, increasing the EU's ability to influence international environmental negotiations.

The fact that cooperation between agents and principals is a crucial characteristic of the EU decision-making process also confirms the added value of extending the principal–agent framework with SI insights. The distribution of preferences and information between the agents and the principals is not as conflicting and in any case not as problematic as traditional principal–agent models assume. Certainly in the authorization and negotiation stages, SI dynamics are an essential feature of the EU decision-making process with regard to international environmental negotiations.

Chapter 7
Conclusions

1. Introduction

This concluding chapter presents the main findings of this study. The conclusions are also presented in Appendix C, together with the corresponding chapter(s) in which their background and explanations can be found. In Section 2, I answer to the research question. Hence, from all the conclusions presented in this chapter, those presented in Section 2 (i.e. conclusions 1a, 1b and 1c) are the key conclusions of this book, since they provide a clear answer to the research question. After having discussed these main conclusions, this section elaborates on the nature of discretion and the interplay of delegation and control in the EU decision-making processes with regard to international negotiations leading to an MEA. It ends with an evaluation of the dilemmas faced by the principals and the agent. The next sections then go back to the various building blocks that were used in this book. Bearing in mind the empirical results of this study, I return to the legal framework (Section 3), the theoretical model (Section 4) and the methodology (Section 5). I present an evaluation of these elements, generally focussing on two questions: what can be learned about them in the light of the empirical data?; and to what extent do the legal, theoretical and methodological propositions hold true?

2. Back to the research question: explaining discretion

2.1. Conditions of the EU negotiator's discretion

2.1.1. Under which conditions does the EU negotiator enjoy (a particular degree of) discretion?
The purpose of this book was to analyse the EU decision-making process with regard to international negotiations leading to an MEA. As such a decision-making process is characterized by delegation of negotiation authority from the member states to an EU negotiator, the analysis was centred around the relation between these two sets of actors. Using principal–agent theory, I conceptualized the delegating member states as 'principals' and the representing EU negotiator as an 'agent'. To comprehend this principal–agent relation and the political dynamics in the decision-making process, I examined the degree of discretion enjoyed by the EU negotiator vis-à-vis the member states when the former negotiates at the international level. Hence, in this research, the EU negotiator's degree of discretion was treated as the outcome variable, i.e. as the political phenomenon I wanted to

explain. As I argued in Chapter 1, a necessary step forward in the literature about the discretion of an EU negotiator in international negotiations was to go beyond single case studies and to examine the conditions determining the EU negotiator's discretion in a comparative way. As a result, my research question, serving as a guideline throughout the various chapters of this book, was formulated as follows: *Which conditions determine the EU negotiator's discretion vis-à-vis the member states during international negotiations leading to a multilateral environmental agreement?*

To answer this question, I studied eight recent EU decision-making processes with regard to international negotiations, each leading to an MEA which is signed by the EC and the member states. Therefore, the following steps were taken. First, the legal boundaries of the process were examined to identify the rules of the game within which such political processes take place (Chapter 2). Second, a theoretical framework was developed (Chapter 3) in which I conceptualized discretion as the sum of the autonomy granted by the member states and the autonomy conquered by the EU negotiator, whereby the granted autonomy was defined as the net result of delegation minus control. On the basis of this theoretical framework, the research design was developed. Third, on the basis of the empirical data (Chapter 4), a QCA procedure was executed in order to systematically compare the cases and to identify the causal paths – i.e. the combinations of conditions – leading to a particular degree of discretion (Chapter 5). Finally, I analysed the empirical data by going back to the variables and to the theoretical framework in order to present a more general picture of the kind of EU decision-making processes studied (Chapter 6).

I distinguished three categories of discretion enjoyed by the EU negotiator during international negotiations leading to an MEA: a *low* degree of discretion, broadly speaking characterized by a large role for the member states at the international level, often negotiating on their own and bypassing the formally appointed EU negotiator; a *moderate* degree of discretion, generally characterized by a situation in which the EU negotiator represents the member states and the latter do not take the floor at the international level, but still quite stringently control the EU negotiator; and, finally, a *high* degree of discretion, characterized by a dominating EU negotiator, acting on the basis of relatively broad instructions and less frequently giving feedback to the member states.

Which conditions explain these categories of discretion? The results of the QCA procedure, as elaborated in Chapter 5, revealed that the answer to the research question is nuanced. These results go beyond explaining the presence and the absence of discretion. Indeed, the distinction between three categories of discretion allowed me to explain why an agent enjoys a *particular degree* of discretion. I deduced three main conclusions out of the explanatory analysis of the EU negotiator's discretion.

[CONCLUSION 1a] A non-compelling external environment explains the absence of discretion for the EU negotiator, while a compelling external environment leads to discretion.

[CONCLUSION 1b] Three causal paths explain a *high* degree of discretion enjoyed by the EU negotiator: [1] institutional density and no politicization; [2] institutional density, diverging member states' preferences and private information for the member states; or [3] politicization and private information for the EU negotiator.

[CONCLUSION 1c] Two causal paths explain a *moderate* degree of discretion enjoyed by the EU negotiator: [1] no private information for the agent and no institutional density; or [2] neither an information benefit for the agents nor for the principals, and politicization.

Starting with conclusion 1a, the results emphasize the decisive role of the compellingness of the external environment for explaining whether an EU negotiator enjoys discretion or not. A non-compelling external environment (i.e. an international negotiation context in which the EU has a relatively large degree of bargaining power and in which the political cost of no agreement is not high for a member state) leads to a *low degree* of discretion for the EU negotiator. In such a negotiation setting (often regional), member states do not prefer to use and to maximize the EU's joint bargaining power. They rather aim to play a large role in the international negotiations on their own and they are able to present their own preferences at the international level. Moreover, as I explained in Chapter 5, neither the member states nor the EU negotiator really need a large degree of discretion to reach an MEA that corresponds to their preferences. Member states can still deploy the control mechanisms without putting at risk the EU's bargaining power or effectiveness. Moreover, there is no external compellingness that the EU negotiator can exploit in order to increase his discretion.

In the same sense, I found that the occurrence of discretion for the EU negotiator – be it a moderate or a high degree, but in any case a particular degree of discretion – can be explained by the presence of a compelling external environment. This means that in an international negotiation context, characterized by a relatively large number of negotiation partners and a relatively limited bargaining power for the EU, an EU negotiator enjoys a certain degree of discretion. More in particular, in such negotiations, member states usually do not take the floor and they grant leeway to their representative. As I theoretically argued in Chapter 3, in such a situation, the political cost of no agreement for the member states increases. The EU negotiator can anticipate on this cost of no agreement and he can exploit the compellingness of the external environment in order to increase his discretion.

As mentioned in conclusions 1b and 1c, the analysis revealed that the EU negotiator's degree of discretion can still considerably vary in a decision-making process characterized by a compelling external environment. Going beyond the black-and-white dichotomy 'discretion/no discretion' allowed me to present a more qualified picture of discretion, as it provided an answer to the question under which conditions an EU negotiator enjoys a high or a moderate degree of discretion. In other words, I could distinguish between, on the one hand, an EU decision-making process in which the EU negotiator is the sole spokesperson for the EU, yet still under a relatively strict control by the member states (i.e. a moderate degree of discretion), and, on the other hand, an EU decision-making process in which the EU negotiator is not only granted exclusive negotiation authority, but in which he is also able to act with a large room for manoeuvre vis-à-vis the member states (i.e. a high degree of discretion).

Three combinations of conditions lead to a *high degree* of discretion:

- a high degree of institutional density and a low degree of politicization;
- a high degree of institutional density and heterogeneous preferences among the member states, which have private information about their fallback positions vis-à-vis the EU negotiator (irrespective of the level of politicization);
- a high degree of politicization and asymmetrically divided information in favour of the EU negotiator about the substance and the developments of the international negotiations.

As I extensively discussed in Chapter 5, in a decision-making processes covered by the first causal path, the risk of too much autonomy for the agent (and the potential costs of delegation) is limited for the principals, because of the cooperative setting and the political insensitivity of the issues. Regarding the second causal path, the degree of discretion can be explained by the empathy of the principals for the agent, and by the possibility for the agent to play the principals against each other and to exploit their private information. In the decision-making processes described in the third causal path, the agent faces a Janus-like role with pressure coming from both the international and the EU level, which he is able to exploit because of the private information he possesses.

A *moderate degree* of discretion can be explained by two causal paths:

- no information benefit for the agent and an institutionally non-dense decision-making context;
- neither an information benefit for the agent nor for the principals in a highly politicized decision-making context.

In the decision-making processes covered by the first causal path, the agent enjoys discretion because the decision-making process is characterized by a compelling external environment, but he is not able to extend his discretion to a high degree because the information about the international negotiations is symmetrically divided between the agent and the principals, and because the principals do not show empathy for the compellingness the agent is confronted with. The second causal path that explains a moderate degree of discretion also relates to decision-making processes where the agent enjoys discretion because of the external compellingness, but where he cannot extend it because all the information is symmetrically divided in the EU and because the principals control the agent as the issues at stake are politically sensitive to them.

2.1.2 Discretion as the result of granted and conquered autonomy

> [CONCLUSION 2] The autonomy granted by the member states is more decisive for the EU negotiator's discretion than the autonomy conquered by the EU negotiator.

As indicated above, I conceptualized discretion as the sum of two components: the autonomy granted by the member states and the autonomy conquered by the EU negotiator. The first component – granted autonomy – is defined as the net result of delegation minus control. In practice, as elaborated in Chapter 6, this first component seems more decisive for the EU negotiator's discretion than the second. What can be learned from this study about the granted and the conquered autonomy?

As I argued in Chapter 3, the inverse of *granted autonomy* is the extent to which the member states control the EU negotiator. Indeed, the more the member states control, the less autonomy they grant. In this sense, I assessed the autonomy granted to the agent by measuring (and then inverting) the control behaviour of the principals, i.e. the extent to which they employed the control mechanisms I distinguished. On the basis of this research, three conclusions can be drawn with regard to the control behaviour of the member states on the EU negotiator, and thus also with regard to the granted autonomy.

> [CONCLUSION 3a] While delegating negotiation authority to the EU negotiator, member states also establish control mechanisms.

> [CONCLUSION 3b] The *ad locum* control mechanisms are the most manifestly used control mechanisms, but it cannot be excluded that the *ex post* ones function in a more latent way.

> [CONCLUSION 3c] The *ad locum* control mechanisms fulfil a mix of control and cooperation functions.

First, referring to conclusion 3a, the central claim of principal–agent theory that delegation and control go hand in hand holds true in the case of EU decision-making processes with regard to international negotiations leading to an MEA. Indeed, to reduce the political costs inherent to delegation, principals control their agent. In the kind of EU decision-making processes studied, I distinguished six control mechanisms that are at the member states' disposal: (1) not authorizing the EU negotiator, and (2) providing a mandate to the EU negotiator (the *ex ante* control mechanisms, employable in the authorization stage); (3) attending the international negotiation settings, and (4) organizing EU coordination meetings (the *ad locum* control mechanisms, employable in the negotiation stage); and (5) not ratifying the MEA, and (6) basing their delegation behaviour in next negotiation rounds on their experiences in the current (the *ex post* control mechanisms, employable in the ratification stage).

Second, referring to conclusion 3b, this research showed that the most manifestly used control mechanisms are the *ad locum* ones. In particular, attending the international negotiations (which implies observing and controlling the negotiation behaviour of the EU negotiator, being able to consult and instruct the EU negotiator on the spot, and even taking the floor at the international level) and organizing EU coordination meetings (which *de facto* forces the EU negotiator to report back on the negotiations and which serves as a forum for the member states to instruct their negotiator) seem to be the most effective control mechanisms (Delreux, 2008). However, as I argued in Chapter 6, it can not be excluded that *ex post* control mechanisms work as well. Although I could not find empirical evidence to support the thesis that the agent negotiates while anticipating the (legal) ratification requirement, it cannot be completely ruled out that this plays a role. In any case, it can be concluded that the *ad locum* control mechanisms are the most manifest expressions of member states restricting the EU negotiator's autonomy. This study did not generate clear evidence for the agent anticipating on the *ex post* control mechanisms.

Third, referring to conclusion 3c, the EU coordination meetings and the member states' attendance of the international negotiation settings do not only have an *ad locum* control function, but also a cooperation function (Delreux, 2009d). As I illustrated in Chapter 6, they constitute a forum for cooperation between the agent and the principals, in which both sets of actors can pool their capabilities in order to strengthen the position of the EU vis-à-vis the international negotiation partners as much as possible. These various manifestations of on the spot coordination can transform the EU into a strong negotiation bloc, which is not dominated by the political – and theoretically expected – tensions between principals and agents, but by the cooperation between all actors involved in the EU process.

> [CONCLUSION 4] Agency slack, in the sense of the EU negotiator conquering more autonomy against the wishes of the member states, is uncommon.

As for the *autonomy conquered* by the EU negotiator, the eight studied decision-making processes pointed out that an EU negotiator conquering autonomy only occurs occasionally. At the very end of the negotiations, the EU negotiator may enjoy more discretion than granted by the member states. In such cases, the agent does not always have the explicit backing of the member states. However, and most importantly, a situation in which the EU negotiator conquers more autonomy *at the cost of the member states*, i.e. going against the wishes of the member states, is very uncommon. Indeed, in the cases studied, I could only find one example of agency slack. Only during the last stage of the Kyoto Protocol negotiations did the troika act at the international level too autonomously and against the position of (some of) the member states. In the other decision-making processes, the member states usually succeeded in keeping their negotiator under control and let him behave as, truly representative. This does not mean that the EU negotiator does not enjoy discretion, but it means that if the EU negotiator enjoys discretion, this is usually wanted by the member states.

2.2. Coping with the dilemmas

In Chapter 1, I presented two dilemmas – one for the principals and one for the agent – that are at the basis of my research question. First, I argued that, if member states aim to exert influence on the substance of an MEA, they face a dilemma between, on the one hand, expressing their own preference in the international negotiations and, on the other hand, aiming to create as much bargaining power as possible. Second, supposing that the EU negotiator is simultaneously involved in two negotiation processes, he has to find a delicate balance between the expectations of both negotiation contexts. What can be learned about these dilemmas on the basis of this research? Or in other words: how do the principals and the agent cope with these dilemmas?

2.2.1. The dilemma of the principals

[CONCLUSION 5a] Member states flesh out their role as principal in a rather pragmatic way.

As for the principals' dilemma, it is clear that practice is not as black-and-white as the dilemma might theoretically suggest. In the EU decision-making processes with regard to international negotiations leading to an MEA, member states do not make radical choices on delegation. Their behaviour is rather pragmatic as they usually neither opt for *full delegation* (the bargaining power strategy) nor for *no delegation* (the own preference strategy). Formally appointing an EU negotiator never seems to be problematic.[1] Once an agent is authorized, the principals generally have a double track to cope with their dilemma.

[CONCLUSION 5b] If a member state is prepared to invest political capabilities in the decision-making process, it can make use of the opportunities of being closely involved in this process and to influence the EU position.

[CONCLUSION 5c] Member states never lose their full degree of negotiation authority.

On the one hand (see conclusion 5b), this study revealed that member states can try to influence as much as possible the position that will be expressed by the EU

---

1   The only exception is the authorization of the Commission in the first stage of the Aarhus Convention negotiations.

negotiator. In this sense, the aim of the principals is to maximize the benefits of both strategies, as they intend to incorporate their own preference in the common EU position, which will then be backed by a large degree of bargaining power and which will be represented by the EU negotiator. Therefore, member states often take up an active position in the EU decision-making process. They have a number of tools at their disposal to do this. First, member states can actively participate in the EU coordination meetings, both on the spot and in Brussels. The principals that take up an active position are able to influence the outcome of the coordination meetings in the direction of their preference. If a member state wants to be influential in the EU decision-making process, it is key to belong to the select circle of principals that are close to the agent. These closely involved member states are the first contact point for the EU negotiator to check things informally, they are part of frequent informal consultation (often by phone or e-mail) in-between the negotiation sessions and they participate in the breakout coordination meetings, which *de facto* adopt a large part of the functions of the regular coordination meeting. Second, the principals can stand for lead country. This not only allows them to play an active role in the international negotiations, but also to prepare and to table EU positions and to have informal contacts with the negotiation partners. Third, member states can assist the EU negotiator during the process by presenting their cooperation, e.g. for the preparation of position papers. Fourth, by sending well-informed and experienced experts to the EU coordination meetings, member states can increase the likelihood of being influential in the EU decision-making process. Principals are thus able to combine the benefits of both strategies, as they have a number of tools at their disposal to influence and to determine the common EU position. However, this does not only require that a member state engages in a cooperative, pragmatic relation with the agent, but also that it is prepared to invest political capabilities, such as expertise, knowledge and personnel, in the process.

On the other hand (see conclusion 5c), authorizing an agent does not automatically imply a definite and absolute loss of negotiation authority for the member states. Even when an EU negotiator is authorized to represent the member states in the international negotiations, the member states still retain the possibility to take the floor – and to express their own preferences – at the international level. This can be considered as a last resort, but it remains a possibility as the international negotiations are mostly held in a UN forum, where (member) states still have full speaking rights. More specifically, principals can first allow the agent to take the floor on their behalf, whereupon they can supplement, refine or even refute the agent's statement. This occurred to a varying extent in the eight cases studied, but it was not always clear whether the intervention by a member state was on its own behalf or on behalf of the EU.

2.2.2 The dilemma of the agent

[CONCLUSION 6] Not only do the principals control the agent, the agent controls the principals as well.

As for the agent's dilemma, my analysis showed that the EU negotiator usually succeeds in finding a balance between the pressure coming from the international level to agree on a certain outcome on the one hand, and the requirement that this outcome should be (politically) approved and (legally) ratified by the member states on the other hand. In the decision-making processes studied, the principals did not reject the agreement reached by the agent. As the EU negotiator is concerned about not losing face vis-à-vis the external negotiation partners, his main objective is to avoid an involuntary defection at the EU level. Therefore, the agent has to keep his principals under control, or at least the agent has to weaken his principals' incentives to control (Delreux, Kerremans, 2010). How does the agent succeed in preventing that the principals deploy their ultimate control mechanism (i.e. non-ratification, also in the sense of rejecting the political agreement at the end of the final negotiation session)? Or in other words: how does the agent control the principals?

Two control strategies for the agent can be distinguished. First, the EU negotiator can make every effort to make the member states experience the compellingness of the external environment. In other words, it is instrumental for the agent to involve the principals in the task he is fulfilling on their behalf. This increases the political cost of no agreement for the principals, which in turn decreases the likelihood of an involuntary defection. In practice, the agent uses the control mechanisms, which are initially established by the principals to control the agent and to decrease the likelihood of agency slack, to keep the principals under control and to avoid an involuntary defection. The frequent (breakout) coordination meetings and debriefings, the member states' attendance at the international negotiation settings and thereby playing the 'mother-in-law role' vis-à-vis the EU negotiator, and the numerous informal consultations with the closely involved member states are also a strategic tool for the agent to assure that the principals would not reject the commitments made vis-à-vis the negotiation partners.

Second, the EU negotiator can play an important role in the agenda-setting of the EU position. As I argued in Chapter 6, the agent usually proposes the EU position to the principals. If the EU negotiator succeeds in being the *first mover* in this process, he can largely determine the borders of his own instructions. This reduces the likelihood of the agent receiving instructions which are unacceptable for the negotiation partners and thus unfeasible at the international level. This, in turn, makes it less likely that the agent is unable to fulfil unrealistic wishes of the principals, which could result in an involuntary defection.

3. Back to the legal framework: EU decision-making and mixed agreements

[CONCLUSION 7] In the authorization and negotiation stages, the EU decision-making process does not completely follow the legal provisions prescribing it *de iure*; in the ratification stage, it does.

My analysis started with a description of the legal framework within which the EU decision-making process with regard to international negotiations leading to an MEA takes place. In Chapter 2, I introduced the complicating factors resulting from the mixed nature of MEAs for the EU decision-making process and I presented the legal rules of the game prescribed by the Treaties on the one hand and the ECJ rulings on the other hand. Chapters 4 and 6 examined how these EU decision-making processes occur in practice. Interpreting the empirical data in the light of the legal framework leads to the conclusion that the empirical reality of EU decision-making processes with regard to international negotiations leading to an MEA does not fully correspond to the legal framework prescribing it, except in the ratification stage.

In the authorization stage, the provisions of article 300 TEC – nowadays article 218 TFEU – are largely followed for issues covered by EC competences: the Commission proposes an authorization decision and a mandate to the member states, which then authorize the Commission to negotiate those issues on their behalf within the boundaries of the existing European legislation, as it is mostly stipulated in the mandate. However, for the issues touching upon national competences, no real authorization (or delegation) decision takes place in practice. Member states do not formally authorize the Presidency or a lead country to negotiate on their behalf. As I argued in Chapter 6, this delegation is regarded as an unconfirmed standard operating procedure. Moreover, member states do not make a clear distinction between their agent for the issues covered by EC competences and the one for the issues dealing with national competences, even when the Commission is granted a mandate. In most cases, the process is thus more fuzzy, as there is often only one EU negotiator, representing the EU as a whole (i.e. the EC and the member states).

In the negotiation stage, the legal framework suggests a clear separation between the Commission negotiating the issues falling under EC competences and the member states (or if they opt to pool their voices, the Presidency) negotiating the issues covered by national competences. In other words, from a legal perspective, the division of competences in the EU should determine the EU negotiation arrangement. However, my analysis showed that the EU negotiation arrangement and the way the EU decision-making process occurs in the negotiation stage are determined by a combination of legal and practical/pragmatic considerations. As I argued in Chapter 6, the legal provisions are only decisive for explaining whether the Commission will play a role as agent. The other aspects of the decision-making process in the negotiation stage are largely determined by practical considerations, driven by a double aspiration: strengthening the EU's position vis-à-vis the negotiation partners at the international level, and involving the interested EU actors in the decision-making process at the EU level. In a lot of cases, the Presidency acted as EU negotiator. However, there are no elements in the legal framework that suggest an agent role for the Presidency. In other words, there is no legal provision (a Treaty article or an ECJ ruling) that prescribes a negotiation role for the Presidency. This is a clear illustration of the difference between the legal framework and the political practice.

In the ratification stage, the EU decision-making process corresponds to what one would expect from it when looking from a legal point of view. In the pre-Lisbon era, the EC ratification followed the standard consultation procedure: the Commission proposes a ratification decision, the European Parliament can give its opinion, and the Council finally decides. Also the national ratifications seem to follow the legally prescribed path, although this study did not examine the political dynamics at the domestic levels.

4. Back to the theoretical framework: evaluating the principal–agent model

4.1. Evaluating the basic model

To theoretically frame the EU decision-making process with regard to international negotiations leading to an MEA, I relied on principal–agent theory. Question is now whether the basic insights of principal–agent theory are in line with the practice of these EU decision-making processes. As I presented in Chapter 3, principal–agent theory basically starts from a three-step reasoning: (1) member states delegate negotiation authority to an EU negotiator because they expect net benefits from it; (2) delegation can also imply political costs, caused by diverging preferences between principals and agent on the one hand, and asymmetrically divided information in favour of the agent on the other hand; (3) to mitigate these political costs, member states create control mechanisms. To what extent does this theoretical reasoning correspond to the empirical reality?

> [CONCLUSION 8a] Delegation to an EU negotiator is rather a standard operating procedure than a well-considered decision (based on a cost-benefit analysis) by the member states.
>
> [CONCLUSION 8b] The cost of delegation for the member states is rather an information cost than a preference cost.

First, referring to conclusion 8a, member states delegate negotiation authority to an EU negotiator. However, as I argued above, this first step is no well-considered decision, resulting from a cost-benefit analysis. As for delegation to the Commission, member states keep to their commitments made in the Treaties and they do usually not contest the Commission's negotiation authority. Moreover, the optional delegation for the issues covered by national competences (i.e. the delegation that is not requested by article 300 TEC/218 TFEU) is considered a standard operating procedure. The rationale behind it seems to be that delegation increases the member states' common bargaining power in the international negotiations, as I suggested in Chapter 3. However, this claim needs to be qualified in a threefold way. First, the member state representatives I interviewed did not explicitly mention the increased bargaining power benefit of delegation. Only when I asked whether bargaining power is a reason for delegation did a large majority of them confirm this. Second, delegation is never exclusive since member states still retain the (theoretical) possibility to take the floor at the international level, although this sometimes would come at a high political cost. Third, delegation to an EU negotiator does not automatically imply that the EU speaks through a single mouth. As I indicated in Chapter 6, presenting a single voice through multiple mouths is often considered as the most effective way of negotiating MEAs for the EU.

Second, referring to conclusion 8b, like principal–agent theory claims, delegation also implies costs for the member states. However, this claim needs to be mitigated: the costs of delegation in the EU decision-making process with regard to international negotiations leading to an MEA are mainly caused by information asymmetry, not by preference heterogeneity between principals and agents.

On the one hand, my findings point out that there is always a risk (or a political cost) that particular information about the substance and the development of the international negotiations will be asymmetrically divided in favour of the agent. As member states usually have the opportunity to attend the international negotiation settings, one may expect that this does not occur. Although the largest part of this cost is indeed eliminated in this way, there are still three reasons why the agent can have private (unshared) information. First, as final deals are usually negotiated in limited negotiation settings (like Friends of the Chair meetings), principals can not rely on first-hand information in such cases. Second, the EU negotiator is mostly regarded as the contact point for the EU by the negotiation partners. This means that the EU negotiator has a comparative advantage vis-à-vis the member states in having an overview of the various positions of the negotiation partners and the possibilities at the international level. Third, in terms of personnel, the EU negotiator usually has the largest team in the EU, which allows the agent to participate in every working and contact group. Hence, the combination of the agent being the only EU representative in the limited negotiation settings, and having the most complete view on the positions of the negotiation partners and on the various negotiation topics, generates a potential cost for the member states.

On the other hand, this analysis showed that in this kind of EU decision-making processes, the agents do usually not have own preferences that are potentially diverging from the principals'. Member states holding the Presidency generally behave in an impartial manner during their six-month term and the Commission does not have separate preferences on international environmental matters, apart from guarding that the provisions in the MEA are not incompatible with existing secondary legislation.

Third, as assumed by principal–agent theory, member states create control mechanisms to keep the agent's discretion within the limits they want. In the context of the studied EU decision-making processes, principals establish *ex ante*, *ad locum* and *ex post* mechanisms. As noted above, the *ex ante* control mechanisms are usually too indistinct and the *ex post* ones too politically costly to exert effective control, and thus to limit discretion. Although the (*ad locum*) control mechanisms certainly contribute to restrict the agent's discretion, they perform two additional functions, which I elaborated above. On the one hand, the consultation between member states and EU negotiator, both in the EU coordination meeting and on the spot, also provides a forum for close cooperation in order to create a strongly performing EU in the international negotiations. On the other hand, these control mechanisms are not only used by the principals to control their agent, but also by the agent to keep his principals under control.

4.2. Evaluating the first extension

To make the principal–agent model applicable to the EU decision-making process with regard to international negotiations leading to an MEA, I adapted it by a threefold extension. How can these extensions be evaluated? Were they useful and did they have an added value to better understand the empirical reality in comparison with a traditional principal–agent model?

> [CONCLUSION 9] When the external environment is compelling, the political cost of no agreement for the principals increases and they reveal their fallback positions. This is, however, a gradual process.

My first extension of the model allowed me to incorporate the compellingness of the external environment. By doing so, I took into account a characteristic of the international negotiation level in order to explain the processes at the EU decision-making level. My research showed that the kind of EU decision-making processes studied are, although to a varying degree, characterized by a compelling external environment, which increases the political cost of no agreement for the member states. The importance (and the validity) of incorporating the external compellingness in the model to understand the EU decision-making process is not only shown by the results of the QCA procedure, which explain the absence and presence of discretion by referring to the compellingness of the external environment (Chapter 5). The various case studies of the decision-making processes (Chapter 4) also revealed that the nature of the international negotiation context determines the relation between principals and agents.

The fact that all studied EU decision-making processes were ultimately approved and ratified seems to indicate that the mechanisms outlined in this extension of the principal–agent model hold true. A compelling external environment increases the political cost of no agreement for a principal. This leads to principals revealing their private information on their fallback positions, which broadens the final win-set of the collective principal. Consequently, the win-set of the collective principal at the end of the process is larger than in the beginning, as theoretically shown in Figures 3.4 and 3.5. However, this extension needs to be qualified. While I theoretically assumed that the member states would only reveal their private information in the ratification stage, this seems to happen more gradually in practice.

As the decision-making processes evolve, principals put their fallback positions on the table. This does not only occur at the end of the decision-making process (i.e. in the ratification stage), but step-by-step during the negotiation stage.

The theoretical exercise made in Section 3.1 of Chapter 3 thus seems to be empirically valid. It contributes to a better understanding of EU decision-making processes with regard to international negotiations leading to an MEA. On the basis of my analysis, I cannot generalize the finding that this causal mechanism and this scope condition play a role in EU decision-making processes dealing with the external dimension of other former first pillar policies. However, I believe that taking into consideration the insight that the external environment matters for explaining the decision-making process at the level below can constitute a fruitful starting point to analyse similar decision-making processes.

4.3. Evaluating the second extension

The second extension's purpose was to include the distinctiveness of mixed agreements in the principal–agent model framing the EU decision-making process. In Section 3.2 of Chapter 3, I argued that the shared nature of environmental competences in the EU – and, as a result, the mixed nature of MEAs – has a double consequence for my theoretical model. Question is now whether these consequences hold true.

> [CONCLUSION 10a] The degree of discretion enjoyed by the agent does not depend on the agent's affiliation, or on whether the agent is a subset of the principals or not.

> [CONCLUSION 10b] The domestic ratification requirement of MEAs does not affect the relation between the member states and the EU negotiator.

On the one hand, negotiating mixed agreements can imply delegation to the Presidency or to a lead country. I claimed that, from a theoretical perspective, it is plausible to expect a lower degree of discretion for the Presidency or a lead country acting as agent than for the Commission taking the agent role. However, my empirical data does not seem to support this proposition (conclusion 10a). By including the condition variable 'agent as subset of the principals' (<agprin>) in my research design, I aimed to measure the effect of the agent being a subset of the principals on the former's discretion. However, no result of the QCA procedure confirmed this thesis. It can thus be concluded that the degree of discretion enjoyed by the EU negotiator does not depend on the affiliation of the actor authorized as agent. In other words, for explaining discretion, the EU negotiator's role (as agent) matters, not his affiliation (Commission, Presidency or lead country). When delegation has taken place, the representation behaviour – and *a fortiori* the discretion – of the Commission, the Presidency or a lead country does not significantly vary.

On the other hand, I presumed that adding the domestic level as an additional ratification requirement would decrease the likelihood of ratification, as the various national parliaments can act as an additional veto player. However, my analysis showed that the domestic ratification requirement does not play a decisive role in the EU decision-making process: a large majority of the MEAs were smoothly ratified at the level of the various member states (conclusion 10b). The domestic ratification requirement for mixed agreements influences neither the relation between principals and agents in the negotiation stage nor the latter's discretion. In other words, the fact that the mixed nature of an MEA adds a third level to the decision-making process does not affect the EU decision-making process. The data of my study showed that extending the principal-model this way is not necessary to understanding the empirical reality of the EU negotiator's discretion, although theoretical considerations suggest differently.

4.4. Evaluating the third extension

The third extension aimed to broaden the principal–agent model for non-RCI dynamics. As traditional principal–agent theory originates RCI and as I did not want to exclude *a priori* that SI would also provide explanatory power for the understanding of EU decision-making processes, I extended the model to an SI account of delegation. To judge whether SI needs to be taken into account in order to fully understand the EU decision-making process, the following two questions need to be answered: (1) is the institutional context of the EU only a forum for strategic (inter)action in which decision-makers behave in an interest-maximizing way?; and (2) are preferences exogenous and readily conflicting when principals and agents engage in a decision-making process within the framework of international environmental negotiations?

[CONCLUSION 11a] The institutional setting of the EU may determine the EU decision-making process, as it prescribes the boundaries of the normative context within which principals and agents behave on the basis of their preferences.

[CONCLUSION 11b] SI offers added value to traditional RCI based principal–agent theory in explaining the decision-making process in the authorization and the negotiation stage.

[CONCLUSION 11c] Not only preferences and institutions matter, personalities matter as well.

[CONCLUSION 11d] National preferences are neither fully exogenously formed nor the pure institutional translation of national interests.

On the role of institutions, my analysis showed that the density of the institutional context of the EU decision-making process is higher than RCI – and traditional principal–agent theory in particular – presumes. Institutions matter, as they have normative power and as they determine the nature of the decision-making process. They prescribe the boundaries of a particular normative context within which principals and agents behave on the basis of their respective preferences (conclusion 11a). This finding is neither surprising nor new, as various studies had already pointed out that EU decision-making cannot be explained from a purely RCI point of view (see Section 3.3.1 of Chapter 3). Hence, I follow the claim of e.g. Lewis, Checkel and Hooghe that the debate between RCI and SI is rather a 'both/and' than an 'either/or' story (Lewis, 2003b; Checkel, 2005b; Hooghe, 2005). However, this book makes an original contribution to the existing literature by incorporating SI insights in a delegation model that explains the relation between principals and agents, and the discretion that results from this relation.

The institutional context thus creates a normative environment. Question is then: which norms create that particular context within which EU actors engage? This research revealed that mainly the striving for compromise and consensus, and trust among the decision-makers matter in the EU decision-making processes with regard to international negotiations leading to an MEA. I could find less evidence for a significant normative influence of mutual responsiveness, diffuse reciprocity (although it played a role among the principals, but not between principals and agents) and avoiding process failure (which is not an objective as such, but rather instrumental to avoid a failure of the international negotiation process).

In Section 3.3.3 of Chapter 3 (and summarized in Figure 3.8), I developed an SI-inspired principal–agent model in order to frame the three stages of the decision-making process from an SI point of view. In Chapter 6, I evaluated this SI model of delegation, concluding that SI provides explanatory power, particularly for the principal–agent relation in the authorization and in the negotiation stage (conclusion 11b). Delegation to the agent in the EU decision-making processes with regard to international negotiations leading to an MEA does usually not occur on the basis of a cost-benefit analysis (RCI). It takes place because it is considered to be an appropriate and legitimate *modus operandi* (SI). Moreover, the decision-making process in the negotiation stage is less conflictual (in terms of maximizing preferences and withholding information) than traditional – RCI-inspired – principal–agent theory assumes. As the relation between principals and agents often takes the form of a partnership in a cooperative problem-solving context, the SI perspective is a useful theoretical tool to understand the way member states and the EU negotiator engage in a decision-making process embedded in a multilateral context. In the ratification stage, finally, the added value of SI is less clear. The ratification of an MEA in the EU is usually regarded as a formality and as a seemingly automatic result of the political commitment made at the end of the negotiations.

This study showed that, besides preferences and institutions, a third EU level determinant shapes the EU decision-making process. Not only preferences and institutions, but also personalities matter (conclusion 11c). Three examples illustrate the decisive role of personalities in the decision-making process. First, the lead country approach and the informal division of labour in the EU is based on the rationale of using the know-how, expertise and capabilities of certain individuals who participate in the decision-making process. Because of the value of their individual characteristics for the EU, these representatives are mobilized to fulfil a task at the international level (e.g. negotiating on behalf of the EU) or at the EU level (e.g. preparing position papers), irrespective of their affiliation. A second example deals with the involvement of a member state in the decision-making process. As I argued above, principals are able to considerably influence the process if they succeed in being closely involved. In practice, this means that they should participate in the breakout coordination meetings, assist the EU negotiator in various tasks, engage in the informal e-mail or phone consultation in-between the negotiation sessions, etc. This research showed that the extent to which a member state is involved in the process does not only depend on the interest a member state has in the issue, but also on the individual capabilities (e.g. experience or expertise) of the particular representative. Third, the influence of an individual representative on a national preference is a final indication of the determining role of personalities in the EU decision-making process.

The second new-institutionalist question dealt with preferences. My research showed that the RCI picture of principals and agents behaving on the basis of exogenous preferences, which maximize their national interests, needs to be qualified (conclusion 11d). On the one hand, preferences can change because of reasons that are not related to national interests or to cost-benefit analyses made at the domestic level. As the principals need to adapt their preferences in function of the developments at the international level, national preferences can also change because of the evolving nature of international negotiations. Moreover, national preferences are endogenous to the institutional context in which the EU decision-making process takes place. The institutional norms that matter in the EU decision-making process (see above) influence the preferences of the actors participating in this process. More in particular, the normative power of trust and consensus and compromise striving can affect the preferences of the principals and the agents.

On the other hand, the assumption that national preferences are the institutional translation of national interests does not always hold. Indeed, in quite a number of cases studied, few national interests were at stake. As I argued in Chapter 6, in these decision-making processes, national preferences are not primarily based on national interests, but on personal know-how and insights of the member state representatives or on the fact that the national preference is to go along with the common EU position, which often originates from existing EC legislation.

5. Back to the method: evaluating QCA

[CONCLUSION 12] QCA may be a fruitful research approach and data analysis method for future studies on EU politics.

In order to identify the (combinations of) conditions determining a particular degree of discretion enjoyed by the EU negotiator, I analysed my data by means of QCA, which is a suitable tool to find patterns in the empirical data in the kind of research presented in this dissertation. Moreover, I argue that QCA may be a fruitful approach for further research in the EU studies. There are only a few studies about EU politics that use QCA (e.g. Bursens, 1999; Kostadinova, 2003; Koenig-Archibugi, 2004; Aus, 2005; Rittberger, Schimmelfennig, 2005; Delreux, 2009a). However, I see three reasons why EU studies can benefit from using QCA, often in combination with other approaches or techniques.

First, QCA allows researchers to analyse and compare data in a systematic way. Many EU studies are limited to one or a small number of cases. Indeed, arguments, conceptual frameworks or theoretical models about EU politics are often only applied to one or a few case(s). Using QCA provides the opportunity to compare cases and to search for scope conditions. In this way, QCA is helpful to go a step further in various areas of the EU studies and to explain the empirical reality of EU politics even in a more complete and comprehensive way. There is a twofold reason for this. On the one hand, QCA results can be generalized towards an adequately defined population, which increases the external validity of the research. In case of this study, the findings hold for the population of recent EU decision-making process with regard to an international negotiation leading to an MEA, which is signed by the EC and the member states. To make this generalization claim, the cases have to be rigorously selected. On the other hand, the results of a QCA procedure can fruitfully be combined with the in-depth knowledge of the empirical cases. Therefore, not only the interpretation of the minimal formulae in the light of the empirical data is an indispensable step in the analysis (see Chapter 5), but QCA should also be accompanied with thorough case studies of the data (see Chapter 4).

Second, and closely related to the first reason, QCA is extremely useful in the so-called 'intermediate-N' situations. In this sense, it is rather surprising that QCA has not frequently been used in EU studies, since a lot of research objects in the EU confront the researcher with an 'intermediate-N' situation. Studies comparing member states, legislative instruments in a particular policy area, negotiation processes (e.g. IGCs, CAP reforms, etc.), Council configurations or Commission DGs, cannot use quantitative statistical methods and could thus benefit from QCA's analytical power.

Third, also the obligatory dichotomization, which is inherent in QCA, can be fruitful for EU studies. Although this step in the procedure is regularly criticized, EU studies can enjoy the benefits of dichotomization. The reason is that EU studies are often confronted with variables the *absolute value* of which is difficult to determine. This was also the case in this study. One of the problems of this kind of research is the conversion of the measurement of e.g. preference distances, information benefits, degree of politicization, degree of institutional density, or even discretion into absolute values. The measuring instruments and operationalization do not allow us to easily code such variables with an absolute value. Moreover, researchers are obliged to appeal to proxies (e.g. interviews or document interpretation) in stead of first-hand information. Dichotomization eliminates the requirement to attribute absolute values to the variables. Using QCA implies that the variables can take *relative values*. On the basis of the researcher's familiarity with the empirical data, a threshold can distinguish between the 'low' and the 'high' values. Consequently, QCA allows researchers to overcome the inherent problem of attributing absolute values to variables by ranking them in a relative way and to define the group of cases that score higher (1 values) than the others (0 values).

Appendices

Appendix A. Interviews

Table A.1   Case-by-case overview of the number of respondents, their affiliation and their role

	CCD	AEWA	KYOTO	ARHUS	PIC	CART	STOCPOP	SEA	Total
Observers		2	1	2		1			6
ECOM		2	1	**1**	**4**	**1**	**2**	**1**	12
MS	6	5	6	4	4	6	7	5	43
AT			*1*		*1*	**1**		*1*	4
BE	**2**		2	2	*1*	*1*	3	*1*	12
DE	*1*	*1*				**1**		*1*	4
DK		*1*							1
ES		*1*							1
FI							*1*		1
FR					*1*		*1*	*1*	3
IT				*1*					1
LU			*1*						1
NL	**2**	*1*	**1**		*1*	**1**	*1*	*1*	8
SE						*1*	*1*		2
UK	*1*	*1*	*1*	*1*		**1**			5
Total	6	9	8	7	8	8	9	6	61

Note: **bold** = agent; *italic* = principal; roman = observer

Table A.2   Overview of the interviews

Id	Case	Affiliation	Role	Date	Place	#min.
1	PIC	Commission	A	24/07/2006	Brussels	75
2	PIC	Commission	A	05/09/2006	Brussels	70
3	CART	Council Secretariat	obs.	07/09/2006	Brussels	120
4	AEWA	Chairman Plenary	obs.	07/09/2006	Phone	60
5	ARHUS	Belgium	P	08/09/2006	Brussels	100
6	PIC	Belgium	P	11/09/2006	Brussels	85
7	STOCPOP	Belgium	P	11/09/2006	Brussels	85
8	CCD	Belgium	P&A	12/09/2006	Ghent	120
9	SEA	Belgium	P	12/09/2006	Brussels	105
10	STOCPOP	Finland	P&A	13/09/2006	Brussels	90
11	AEWA	Germany	P	14/09/2006	Bonn	120
12	CART	Belgium	P	15/09/2006	Brussels	60
13	STOCPOP	Commission	A	19/09/2006	Brussels	75
14	PIC	Commission	A	20/09/2006	Brussels	105
15	PIC	Commission	A	20/09/2006	Brussels	105
16	ARHUS	Belgium	P	20/09/2006	Brussels	75
17	STOCPOP	Belgium	P	21/09/2006	Brussels	105
18	AEWA	Adviser Commission	obs.	21/09/2006	Brussels	120
19	AEWA	Adviser Commission	obs.	21/09/2006	Brussels	120
20	SEA	Commission	A	26/09/2006	Brussels	90
21	ARHUS	UK	P	27/09/2006	London	90
22	STOCPOP	Sweden	P	28/09/2006	Brussels	90
23	KYOTO	Belgium	P	28/09/2006	Brussels	120
24	SEA	France	P	29/09/2006	Paris	120
25	PIC	France	P	04/10/2006	Paris	75
26	STOCPOP	France	P&A	18/10/2006	Brussels	105
27	SEA	The Netherlands	P	19/10/2006	The Hague	90
28	AEWA	Chairman Plenary	obs.	25/10/2006	Jambes	60
29	AEWA	The Netherlands	P	30/10/2006	The Hague	135
30	STOCPOP	The Netherlands	P	30/10/2006	The Hague	150

Id	Case	Affiliation	Role	Date	Place	#min.
31	KYOTO	Belgium	P	06/11/2006	Brussels	135
32	STOCPOP	Belgium	P	07/11/2006	Brussels	90
33	CART	Commission	A	07/11/2006	Phone	105
34	KYOTO	Luxembourg	P&A	08/11/2006	Luxembourg	70
35	KYOTO	Council Secretariat	obs.	10/11/2006	Brussels	120
36	ARHUS	Italy	P	14/11/2006	Phone	70
37	CART	The Netherlands	P&A	15/11/2006	Lasne-Ghent-Lasne	120
38	CCD	UK	P	16/11/2006	London	100
39	KYOTO	Commission	obs.	17/11/2006	Brussels	105
40	CCD	Germany	P	20/11/2006	Bonn	120
41	ARHUS	Commission	A	22/11/2006	Brussels	90
42	PIC	The Netherlands	P	23/11/2006	The Hague	90
43	AEWA	Spain	P	24/11/2006	Phone	30
44	CART	Austria	P&A	27/11/2006	Phone	60
45	KYOTO	Austria	P	27/11/2006	Brussels	70
46	SEA	Austria	P	04/12/2006	Brussels	80
47	SEA	Germany	P	04/12/2006	Brussels	90
48	AEWA	UK	P	04/12/2006	Phone	60
49	ARHUS	Chairman Plenary	obs.	06/12/2006	Oegstgeest	100
50	AEWA	Denmark	P	07/12/2006	Phone	60
51	CCD	Belgium	P&A	07/12/2006	Brussels	130
52	KYOTO	The Netherlands	A	11/12/2006	Voorburg	90
53	PIC	Austria	P	12/12/2006	Brussels	150
54	STOCPOP	Commission	A	13/12/2006	Brussels	110
55	CCD	The Netherlands	P	20/12/2006	The Hague	80
56	ARHUS	NGO Coalition	obs.	21/12/2006	Phone	100
57	CART	UK	P&A	10/01/2007	London	90
58	CART	Sweden	P	15/01/2007	Phone	60
59	CCD	The Netherlands	P	18/01/2007	The Hague	80
60	KYOTO	UK	P&A	19/01/2007	London	75
61	CART	Germany	P&A	23/01/2007	Brussels	120

Note: 'P': principal; 'A': agent; 'obs.': observer

Appendix B.

Table B.1   International negotiation sessions leading to the selected MEAs

CCD (Intergovernmental Negotiating Committee on Desertification)			
1	INCD 1	24/05/1993-03/06/1993	Nairobi, Kenya
2	INCD 2	13/09/1993-24/09/1993	Geneva, Switzerland
3	INCD 3	17/01/1994-28/01/1994	New York, US
4	INCD 4	21/03/1994-31/03/1994	Geneva, Switzerland
5	INCD 5	06/06/1994-17/06/1994	Paris, France
s	signing	14/10/1994-15/10/1994	Paris, France
AEWA			
	expert meetings	since 1985	
1	informal negotiation meeting	12/06/1994-14/06/1994	Nairobi, Kenya
2	formal negotiation meeting	12/06/1995-17/06/1995	The Hague, The Netherlands
s	signing	16/06/1995	The Hague, The Netherlands
KYOTO (Subsidiary Bodies; Conference of the Parties)			
1	SB 5	25/02/1997-07/03/1997	Bonn, Germany
2	SB 6	28/07/1997-07/08/1997	Bonn, Germany
3	SB 7	20/10/1997-31/10/1997	Bonn, Germany
4	COP 3	01/12/1997-10/12/1997	Kyoto, Japan
s	signing	11/12/1997	Kyoto, Japan
ARHUS (Working Group for the preparation of a draft convention)			
1	WG 1	17/06/1996-19/06/1996	Geneva, Switzerland
2	WG 2	30/10/1996-01/11/1996	Geneva, Switzerland
3	WG 3	11/12/1996-13/12/1996	Geneva, Switzerland

ARHUS (Working Group for the preparation of a draft convention) *(cont.)*			
4	WG 4	19/02/1997-21/02/1997	Geneva, Switzerland
5	WG 5	18/06/1997-20/06/1997	Geneva, Switzerland
6	WG 6	09/07/1997-11/07/1997	Geneva, Switzerland
7	WG 7	29/09/1997-03/10/1997	Geneva, Switzerland
8	WG 8	01/12/1997-05/12/1997	Rome, Italy
9	WG 9	12/01/1998-16/01/1998	Geneva, Switzerland
10	WG 10	03/03/1998-06/03/1998	Geneva, Switzerland
s	signing	25/06/1998	Aarhus, Denmark
PIC (Intergovernmental Negotiating Committee)			
1	INC 1	11/03/1996-15/03/1996	Brussels, Belgium
2	INC 2	16/09/1996-20/09/1996	Nairobi, Kenya
3	INC 3	26/05/1997-30/05/1997	Geneva, Switzerland
4	INC 4	20/10/1997-24/10/1997	Rome, Italy
5	INC 5	09/03/1998-14/03/1998	Brussels, Belgium
s	signing	10/09/1998	Rotterdam, The Netherlands
CART (Open-ended Ad Hoc Working Group on Biosafety; Extraordinary Meeting of the Conference of the Parties; Informal Consultation (IC) on the process to resume the Extraordinary Meeting of COP)			
1	BSWG 1	22/07/1996-26/07/1996	Aarhus, Denmark
2	BSWG 2	12/05/1997-16/05/1997	Montreal, Canada
3	BSWG 3	13/10/1997-17/10/1997	Montreal, Canada
4	BSWG 4	05/02/1998-13/02/1998	Montreal, Canada
5	BSWG 5	17/08/1998-28/08/1998	Montreal, Canada
6	BSWG 6	14/02/1999-19/02/1999	Cartagena, Columbia
7	ExCOP	22/02/1999-23/02/1999	Cartagena, Columbia
8	IC	15/09/1999-19/09/1999	Vienna, Austria
9	ExCOP-bis	24/01/2000-29/01/2000	Montreal, Canada
s	signing	24/05/2000	Nairobi, Kenya

STOCPOP (Intergovernmental Negotiating Committee)			
1	INC 1	29/06/1998-03/07/1998	Montreal, Canada
2	INC 2	25/01/1999-29/01/1999	Nairobi, Kenya
3	INC 3	06/09/1999-11/09/1999	Geneva, Switzerland
4	INC 4	20/03/2000-25/03/2000	Bonn, Germany
5	INC 5	04/12/2000-09/12/2000	Johannesburg, South Africa
s	signing	22/05/2001	Stockholm, Sweden
SEA (open-ended Ad Hoc Working Group on the SEA Protocol)			
1	WG 1	14/05/2001-16/05/2001	Geneva, Switzerland
2	WG 2	26/09/2001-28/09/2001	Geneva, Switzerland
3	WG 3	21/11/2001-23/11/2001	Orvieto, Italy
4	WG 4	11/02/2002-13/02/2002	Warsaw, Poland
5	WG 5	06/05/2002-08/05/2002	Oslo, Norway
6	WG 6	23/09/2002-27/09/2002	Ohrid, FYROM
7	WG 7	18/11/2002-22/11/2002	Geneva, Switzerland
8	WG 8	30/01/2003	Geneva, Switzerland
s	signing	21/05/2003	Kiev, Ukraine

Appendix C.

Table C.1 Overview of the conclusions and their corresponding chapters

		Chapter 1. Introduction	Chapter 2. The EU and mixed agreements	Chapter 3. Member states as principals, EU negotiators as agents	Chapter 4. The EU as a negotiator in eight international environmental negotiations	Chapter 5. Explaining the EU negotiator's discretion	Chapter 6. Unravelling the EU decision-making process
1a.	A non-compelling external environment explains the absence of discretion for the EU negotiator, while a compelling external environment leads to discretion.	•		•		•	
1b.	Three causal paths ([1] institutional density and no politicization; [2] institutional density, diverging member states' preferences and private information for the member states; or [3] politicization and private information for the EU negotiator) explain a *high* degree of discretion enjoyed by the EU negotiator.	•		•		•	
1c.	Two causal paths ([1] no private information for the agent and no institutional density; or [2] neither an information benefit for the agents nor for the principals, and politicization) explain a *moderate* degree of discretion enjoyed by the EU negotiator.	•		•		•	

		Chapter 1. Introduction	Chapter 2. The EU and mixed agreements	Chapter 3. Member states as principals, EU negotiators as agents	Chapter 4. The EU as a negotiator in eight international environmental negotiations	Chapter 5. Explaining the EU negotiator's discretion	Chapter 6. Unravelling the EU decision-making process
2.	The autonomy granted by the member states is more decisive for the EU negotiator's discretion than the autonomy conquered by the EU negotiator.			•	•		•
3a.	While delegating negotiation authority to the EU negotiator, member states also establish control mechanisms.			•	•		•
3b.	The *ad locum* control mechanisms are the most manifestly used control mechanisms, but it cannot be excluded that the *ex post* ones function in a more latent way.			•	•		•
3c.	The *ad locum* control mechanisms fulfil a mix of control and cooperation functions.			•	•		•
4.	Agency slack, in the sense of the EU negotiator conquering more autonomy against the wishes of the member states, is uncommon.			•	•		•
5a.	Member states flesh out their role as principal in a rather pragmatic way.	•			•		•

		Chapter 1. Introduction	Chapter 2. The EU and mixed agreements	Chapter 3. Member states as principals, EU negotiators as agents	Chapter 4. The EU as a negotiator in eight international environmental negotiations	Chapter 5. Explaining the EU negotiator's discretion	Chapter 6. Unravelling the EU decision-making process
5b.	If a member state is prepared to invest political capabilities in the decision-making process, it can make use of the opportunities of being closely involved in this process and to influence the EU position.	•			•		•
5c.	Member states never lose their full degree of negotiation authority.	•			•		•
6.	Not only do the principals control the agent, the agent controls the principals as well.	•			•		•
7.	In the authorization and negotiation stages, the EU decision-making process does not completely follow the legal provisions prescribing it *de iure*; in the ratification stage, it does.		•		•		•
8a.	Delegation to an EU negotiator is rather a standard operating procedure than a well-considered decision (based on a cost-benefit analysis) by the member states.			•	•		•

		Chapter 1. Introduction	Chapter 2. The EU and mixed agreements	Chapter 3. Member states as principals, EU negotiators as agents	Chapter 4. The EU as a negotiator in eight international environmental negotiations	Chapter 5. Explaining the EU negotiator's discretion	Chapter 6. Unravelling the EU decision-making process
8b.	The cost of delegation for the member states is rather an information cost than a preference cost.			•	•		•
9.	When the external environment is compelling, the political cost of no agreement for the principals increases and they reveal their fallback positions. This is, however, a gradual process.			•	•	•	•
10a.	The degree of discretion enjoyed by the agent does not depend on the agent's affiliation, or on whether the agent is a subset of the principals or not.			•	•	•	•
10b.	The domestic ratification requirement of MEAs does not affect the relation between the member states and the EU negotiator.			•	•		•
11a.	The institutional setting of the EU may determine the EU decision-making process, as it prescribes the boundaries of the normative context within which principals and agents behave on the basis of their preferences.			•	•		•

		Chapter 1. Introduction	Chapter 2. The EU and mixed agreements	Chapter 3. Member states as principals, EU negotiators as agents	Chapter 4. The EU as a negotiator in eight international environmental negotiations	Chapter 5. Explaining the EU negotiator's discretion	Chapter 6. Unravelling the EU decision-making process
11b.	SI offers added value to traditional RCI-based principal–agent theory in explaining the decision-making process in the authorization and the negotiation stage.			•	•		•
11c.	Not only preferences and institutions matter, personalities matter as well.			•	•		•
11d.	National preferences are neither fully exogenously formed nor the pure institutional translation of national interests.			•	•		•
12.	QCA may be a fruitful research approach and data analysis method for future studies on EU politics.					•	

Bibliography

Primary sources

Council of Ministers, *Council conclusions of the 1873rd Council meeting*, 06/10/1995, 10208/95 (Press 275).

Council of Ministers, *Council conclusions of the 1939th Council meeting*, 25/06/1996, 8518/96 (Press 188).

Council of Ministers, *Council conclusions of the 1990th Council meeting*, 03/03/1997, 6309/97 (Press 60).

Council of Ministers, *Council conclusions of the 2017th Council meeting*, 19/06/1997, 9132/97 (Press 204).

Council of Ministers, *Council conclusions of the 2033rd Council meeting*, 16/10/1997, 11332/97 (Press 296).

Council of Ministers, *Council conclusions of the 2106th Council meeting*, 16/06/1998, 09402/98 (Press 205).

Council of Ministers, *Council conclusions of the 2108th Council meeting*, 17/06/1998, 09551/98 (Press 207).

Council of Ministers, *Council conclusions of the 2235th Council meeting*, 13/12/1999, 13854/99 (Press 409).

Council of Ministers, *Council conclusions of the 2302nd Council meeting*, 07/11/2000, 12924/00 (Press 416).

European Commission, *Communication from the Commission to the Council, the European Parliament, the Economic and Social Committee and the Committee of the Regions. Climate Change – The EU Approach for Kyoto*, COM(1997) 481, 01/10/1997.

European Commission, *Proposal for a Council Decision concerning the approval, on behalf of the European Community, of the Kyoto Protocol to the United Nations Framework Convention on Climate Change and the joint fulfilment of commitments thereunder*, COM(2001) 579, 23/10/2001.

European Commission, *Proposal for a Council decision approving, on behalf of the European Community, the Rotterdam Convention on the Prior Informed Consent Procedure for certain hazardous chemicals and pesticides in international trade*, COM(2001) 802, 24/01/2002.

European Commission, *Proposal for a Council decision concerning the conclusion, on behalf of the Community, of the Cartagena Protocol on Biosafety*, COM(2002) 62, 13/03/2002.

European Commission, *Proposal for a Council decision concerning the conclusion, on behalf of the European Community, of the Stockholm Convention on Persistent Organic Pollutants*, COM(2003) 331, 12/06/2003.

European Commission, *Proposal for a Regulation of the European Parliament and of the Council on the application of the provisions of the Århus Convention on Access to Information, Public Participation in Decision-making and Access to Justice in Environmental Matters to EC institutions and bodies*, COM(2003) 622, 24/10/2003.

European Commission, *Proposal for a Directive of the European Parliament and of the Council on access to justice in environmental matters*, COM(2003) 624, 24/10/2003.

European Commission, *Proposal for a Council Decision on the conclusion, on behalf of the European Community, of the Convention on access to information, public participation in decision making and access to justice regarding environmental matters*, COM(2003) 625, 24/10/2003.

European Commission, *Proposal for a Council decision on the conclusion by the European Community of the Agreement on the Conservation of African-Eurasian Migratory Waterbirds*, COM(2004) 531, 03/08/2004.

European Community, Council Regulation of 23 July 1992 concerning the export and import of certain dangerous chemicals in *Official Journal*, 2455/92/EEC, L251, 29/08/1992, 13-22.

European Community, Council Directive of 16 September 1996 on the disposal of polychlorinated biphenyls and polychlorinated terphenyls (PCB/PCT). *Official Journal*, L243, 96/59/EC, 24/09/1996, 31-35.

European Community, Council Directive of 24 September 1996 concerning integrated pollution prevention and control. *Official Journal*, L257, 96/61/EC, 10/10/1996, 26-40.

European Community, Council Directive of 3 March 1997 amending Directive 85/337/EEC on the assessment of the effects of certain public and private projects on the environment. *Official Journal*, L73, 97/11/EC, 14/03/1997, 5-15.

European Community, Council Decision of 9 March 1998 on the conclusion, on behalf of the European Community, of the United Nations Convention to combat desertification in countries seriously affected by drought and/or desertification, particularly in Africa. *Official Journal*, L83, 98/216/EC, 19/03/1998, 1-2.

European Community, Directive of the European Parliament and of the Council of 27 June 2001 on the assessment of the effects of certain plans and programmes on the environment. *Official Journal*, L197, 2001/42/EC, 21/07/2001, 30-37.

European Community, Council Decision of 25 April 2002 concerning the approval, on behalf of the European Community, of the Kyoto Protocol to the United Nations Framework Convention on Climate Change and the joint fulfilment of commitments thereunder. *Official Journal*, L130, 2002/358/EC, 15/05/2002, 1-3.

European Community, Council Decision of 25 June 2002 concerning the conclusion, on behalf of the European Community, of the Cartagena Protocol on Biosafety. *Official Journal*, L201, 2002/628/EC, 31/07/2002, 48-49.

European Community, Directive of the European Parliament and of the Council of 28 January 2003 on public access to environmental information and repealing Council Directive 90/313/EEC. *Official Journal*, L41, 2003/4/EC, 14/02/2003, 26-32.

European Community, Council Decision of 19 December 2002 concerning the approval, on behalf of the European Community, of the Rotterdam Convention on the Prior Informed Consent Procedure for certain hazardous chemicals and pesticides in international trade. *Official Journal*, L63, 2003/106/EC, 06/03/2003, 27-47.

European Community, Directive of the European Parliament and of the Council of 26 May 2003 providing for public participation in respect of the drawing up of certain plans and programmes relating to the environment and amending with regard to public participation and access to justice Council Directives 85/337/ EEC and 96/61/EC. *Official Journal*, L156, 2003/35/EC, 25/06/2003, 17-25.

European Community, Council Decision of 17 February 2005 on the conclusion, on behalf of the European Community, of the Convention on access to information, public participation in decision-making and access to justice in environmental matters. *Official Journal*, L124, 2005/370/EC, 17/05/2005, 1-3.

European Community, Council Decision of 14 October 2004 concerning the conclusion, on behalf of the European Community, of the Stockholm Convention on Persistent Organic Pollutants. *Official Journal*, L209, 2006/507/ EC, 31/07/2006, 1-2.

European Community, Regulation of the European Parliament and of the Council of 6 September 2006 on the application of the provisions of the Aarhus Convention on Access to Information, Public Participation in Decision-making and Access to Justice in Environmental Matters to Community institutions and bodies. *Official Journal*, L264, 2006/1367/EC, 25/09/2006, 13-19.

European Community, Council Decision of 25 September 2006 on the conclusion, on behalf of the European Community, of the Rotterdam Convention on the Prior Informed Consent Procedure for certain hazardous chemicals and pesticides in international trade. *Official Journal*, L299, 2006/730/EC, 28/10/2006, 23-25.

European Community, Council Decision of 18 July 2005 on the conclusion on behalf of the European Community of the Agreement on the Conservation of African-Eurasian Migratory Waterbirds. *Official Journal*, L345, 2006/871/EC, 08/12/2006, 24-25.

European Community, Council Decision of 20 October 2008 on the approval, on behalf of the European Community, of the Protocol on Strategic Environmental Assessment to the 1991 UN/ECE Espoo Convention on Environmental Impact Assessment in a Transboundary Context. *Official Journal*, L308, 2008/871/ EC, 19/11/2008, 33-34.

European Court of Justice, *Opinion 2/00 of the Court*, 06/12/2001.

European Economic Community, Council Directive of 21 December 1978 prohibiting the placing on the market and use of plant protection products containing certain active substances. *Official Journal*, L33, 79/117/EEC, 08/02/1979, 36-40.

European Economic Community, Council Directive of 2 April 1979 on the conservation of wild birds. *Official Journal*, L10, 379/409/EEC, 25/04/1979, 1-18.

European Economic Community, Council Directive of 27 June 1985 on the assessment of the effects of certain public and private projects on the environment. *Official Journal*, L175, 85/337/EEC, 05/07/1985, 40-48.

European Economic Community, Council Directive of 23 April 1990 on the contained use of genetically modified micro-organisms. *Official Journal*, L117, 90/219/EEC, 08/05/1990, 1-14.

European Economic Community, Council Directive of 23 April 1990 on the deliberate release into the environment of genetically modified organisms. *Official Journal*, L117, 90/220/EEC, 08/05/1990, 15-27.

European Economic Community, Council Directive of 7 June 1990 on the freedom of access to information on the environment. *Official Journal*, L158, 90/313/EEC, 23/06/1990, 56-58.

European Economic Community, Council Directive of 21 May 1992 on the conservation of natural habitats and of wild fauna and flora. *Official Journal*, L206, 92/43/EEC, 22/07/1992, 7-50.

European Parliament, *Report on the proposal for a Council decision approving, on behalf of the European Community, the Rotterdam Convention on the Prior Informed Consent Procedure for certain hazardous chemicals and pesticides in international trade*, A5-0290/2002, 11/09/2002.

Secondary sources

Aarebrot F., Bakka P. (1997), Die Vergleichende Methode in der Politikwissenschaft, in Berg-Schlosser D., Müller-Rommel F. (eds.), *Vergleichende Politikwissenschaft*, Opladen, Leske & Burdrich, 49-66.

Aggarwal V., Fogarty E. (2004), Explaining Trends in EU Interregionalism, in Aggarwal V., Fogarty E. (eds.), *EU Trade Strategies. Between Regionalism and Globalism*, New York, Palgrave Macmillan, 207-240.

Andrée P. (2005), The Cartagena Protocol on Biosafety and Shifts in the Discourse of Precaution. *Global Environmental Politics*, 5(4), 25-46.

Arksey H., Knight P. (1999), *Interviewing for Social Scientists*, London, Sage Publications.

Arregui J., Stokman F., Thomson R. (2004), Bargaining in the European Union and Shifts in Actors' Policy Positions. *European Union Politics*, 5(1), 47-72.

Aspinwall M., Schneider G. (2001), Institutional research on the European Union: mapping the field, in Aspinwall M., Schneider G. (eds.), *The rules of integration. Institutionalist approaches to the study of Europe*, Manchester, Manchester University Press, 1-18.

Aus J. (2005), *Conjunctural Causation in Comparative Case-Oriented Research. Exploring the Scope Conditions of Rationalist and Institutionalist Causal Mechanisms*, Oslo, ARENA, Working Paper WP 05/25, http://www.arena.uio. no/publications/working-papers2005/papers/wp05_28.pdf.

Axelrod R. (1981), The Emergence of Cooperation among Egoists. *The American Political Science Review*, 75(2), 306-318.

Bail C., Decaestecker J., Jørgensen M. (2002), European Union, in Bail C., Falkner R., Marquard H. (eds.), *The Cartagena Protocol on Biosafety. Reconciling Trade in Biotechnology with Environment & Development?*, London, Royal Institute of International Affairs, 166-185.

Baldwin M. (2006), EU Trade Politics – Heaven or Hell?. *Journal of European Public Policy*, 13(6), 926-942.

Ballmann A., Epstein D., O'Halloran S. (2002), Delegation, Comitology, and the Separation of Powers in the European Union. *International Organization*, 56(3), 551-574.

Bendor J., Glazer A., Hammond T. (2001), Theories of Delegation. *Annual Review of Political Science*, 4(–), 235-269.

Berg-Schlosser D., De Meur G., Rihoux B., Ragin C. (2009), Qualitative Comparative Analysis (QCA) as an Approach, in Rihoux B., Ragin C. (eds.), *Configurational Comparative Methods. Qualitative Comparative Analysis (QCA) and Related Techniques*, London, Sage Publications, 1-18.

Betz N. (2008), *Mixed Agreements – EC and EU*, paper presented at the Garnet Conference 'The European Union in International Affairs', Brussels.

Bevilacqua D. (2007), The International Regulation of Genetically Modified Organisms: Uncertainty, Fragmentation, and Precaution. *European Environmental Law Review*, 16(12), 314-336.

Beyers J. (2005), Multiple Embeddedness and Socialization in Europe: The Case of Council Officials. *International Organization*, 59(4), 899-936.

Billiet S. (2006), From GATT to the WTO: The Internal Struggle for External Competences in the EU. *Journal of Common Market Studies*, 44(5), 899-919.

Billiet S. (2009), Principal-agent analysis and the study of the EU: What about the EC's external relations?. *Comparative European Politics*, 7(4), 435-454.

Bretherton C., Vogler J. (2000), The European Union as Trade Actor and Environmental Activist: Contradictory Roles?. *Journal of Economic Integration*, 15(2), 163-194.

Bretherton C., Vogler J. (2003), *The European Union as a Global Actor*, London, Routledge.

Bueno de Mesquita B. (1994), Political Forecasting: An Expected Utility Method, in Bueno de Mesquita B., Stokman F. (eds.), *European Community Decision Making. Models, Applications, and Comparisons*, New Haven, Yale University Press, 71-104.

Bueno de Mesquita B. (2004), Decision-Making Models, Rigor and New Puzzles. *European Union Politics*, 5(1), 125-138.

Burgiel S. (2002), The Cartagena Protocol on Biosafety: Taking Steps from Negotiation to Implementation. *Review of European Community and International Environmental Law*, 11(1), 53-61.

Bursens P. (1999), *Impact van instituties op beleidsvorming. Een institutioneel perspectief op besluitvorming in de communautaire pijler van de Europese Unie*, Antwerpen, Universiteit Antwerpen, Departement Politieke en Sociale Wetenschappen.

Cameron F. (2004), After Iraq: The EU and Global Governance. *Global Governance*, 10(2), 157-163.

Chagas V. (2003), The European Union bubble: Differentiation in the assignment of the greenhouse gas emission targets. *Journal of European Integration*, 25(2), 151-163.

Checkel J. (2001), Why Comply? Social Learning and European Identity Change. *International Organization*, 55(3), 553-588.

Checkel J. (2004), Social constructivisms in global and European politics: a review essay. *Review of International Studies*, 30(2), 229-244.

Checkel J. (2005a), *It's the Process Stupid! Process Tracing in the Study of European and International Politics*, Oslo, ARENA, Working Paper WP 05/26, http://www.arena.uio.no/publications/ working-papers2005/papers/ wp05_26.pdf.

Checkel J. (2005b), International Institutions and Socialization in Europe: Introduction and Framework. *International Organization*, 59(4), 801-826.

Checkel J. (2006), Tracing Causal Mechanisms. *International Studies Review*, 8(2): 362-370.

Christensen T., Lægreid P. (2007), Regulatory Agencies. The Challenges of Balancing Agency Autonomy and Political Control. *Governance*, 20(3), 499-520.

Clément C. (2004), Un modèle commun d'effondrement de l'Etat? Une AQQC du Liban, de la Somalie et de l'ex-Yougoslavie. *Revue Internationale de Politique Comparée*, 11(1), 35-50.

Cœuré B., Pisani-Ferry J. (2003), *One Market, One Voice? European Arrangements in International Economic Relations*, paper prepared for the conference on "New Institutions for a New Europe", Vienna.

Coleman W., Tangermann S. (1999), The 1992 CAP reform, the Uruguay Round and the Commission: conceptualizing linked policy games. *Journal of Common Market Studies*, 37(3), 385-404.

Collinson S. (1999), 'Issue-systems', 'multi-level games' and the analysis of the EU's external commercial and associated policies: a research agenda. *Journal of European Public Policy*, 6(2), 206-224.

Costa O. (2006), *Why does the EU lead international climate negotiations? Actors, alliances and institutions*, paper presented at the 3rd Pan-European ECPR Conference on EU Politics, Istanbul.

Cronqvist L. (2003), *Using Multi-Value Logic Synthesis in Social Science*, ECPR General Conference, Marburg.

Cutcher-Gershenfeld J., Watkins M. (1999), Toward a Theory of Representation in Negotiation, in Mnookin R., Susskind L. (eds.), *Negotiating on behalf of others*, London, Sage Publications, 23-51.

Damro C. (2006), EU-UN Environmental Relations: Shared Competence and Effective Multilateralism, in Smith K., Laatikainen K. (eds.), *Intersecting Multilateralisms. The European Union and the United Nations*, New York, Palgrave Macmillan, 175-192.

Damro C. (2007), EU Delegation and Agency in International Trade Negotiations: A Cautionary Comparison. *Journal of Common Market Studies*, 45(4), 883-903.

Damro C., Hardie I., MacKenzie D. (2008), The EU and Climate Change Policy: Law, Politics and Prominance at Difference Levels. *Journal of Contemporary European Research*, 4(3), 179-192.

Damro C., Luaces P. (2001), *Emissions trading – a Kyoto Nepi: From EU Resistance to Union Innovation*, paper presented at the ECPR Joint Sessions, Grenoble.

Dashwood A., Heliskoski J. (2000), The Classic Authorities Revisited, in Dashwood A., Hillion C. (eds.), *The General Law of EC External Relations*, London, Sweet & Maxwell, 3-19.

De Bièvre D., Dür A. (2005), Constituency Interests and Delegation in European and American Trade Policy. *Comparative Political Studies*, 38(10), 1271-1296.

De Meur G., Rihoux B. (2002), *L'Analyse Quali-Quantitative Comparée (AQQC-QCA). Approche, techniques et applications en sciences humaines*, Louvain-la-Neuve, Academia Bruylant.

Degrand-Guillaud A. (2009), Characteristics of and Recommendations for EU Coordination at the UN. *European Foreign Affairs Review*, 14(4), 607-622.

Delreux T. (2006), The European Union in international environmental negotiations: a legal perspective on the internal decision-making process. *International Environmental Agreements*, 6(3), 231-248.

Delreux T. (2008), The EU as a negotiator in multilateral chemicals negotiations: multiple principals, different agents. *Journal of European Public Policy*, 15(7), 1069-1086.

Delreux T. (2009a), The EU negotiates multilateral environmental agreements: explaining the agent's discretion. *Journal of European Public Policy*, 16(5), 719-737.

Delreux T. (2009b), The EU in Environmental Negotiations in UNECE: An Analysis of its Role in the Aarhus Convention and the SEA Protocol Negotiations. *Review of European Community and International Environmental Law*, 18(3), 328-337.

Delreux T. (2009c), The European Union in International Environmental Negotiations: an Analysis of the Stockholm Convention Negotiations. *Environmental Policy and Governance*, 19(1), 21-31.

Delreux T. (2009d), Cooperation and Control in the European Union. The Case of the European Union as International Environmental Negotiator. *Cooperation and Conflict*, 44(2), 189-208.

Delreux T., Hesters D. (2010), *Solving contradictory simplifying assumptions in QCA: presentation of a new best practice*, Louvain-la-Neuve, Compasss, http://www.compasss.org/files/WPfiles/Delreux2010.pdf.

Delreux T., Kerremans B. (2010), How Agents Weaken their Principals' Incentives to Control: The Case of EU Negotiations and EU Member States in Multilateral Negotiations. *Journal of European Integration*, 32(4), 247-264.

Dessai S., Schipper E. (2003), The Marrakech Accords to the Kyoto Protocol: analysis and future prospects. *Global Environmental Change*, 13(2), 149-153.

Druckman J., Lupia A. (2000), Preference Formation. *Annual Review of Political Science*, 3(–), 1-24.

Dür A. (2006), Assessing the EU's role in international trade negotiations. *European Political Science*, 5(4), 362-376.

Dür A., Zimmermann H. (2007), Introduction: The EU in International Trade Negotiations. *Journal of Common Market Studies*, 45(4), 771-787.

Dutzler B. (2002), The Representation of the EU and the Member States in International Organisations – General Aspects, Griller S., Weidel B. (eds.), *External Economic Relations and Foreign Policy in the European Union*, Vienna, Springer-Verlag, 151-189.

Eeckhout P. (2004), *External Relations of the European Union*, New York, Oxford University Press.

Elgström O. (2006), *Leader or Foot-Dragger? Perceptions of the European Union in Multilateral International Negotiations*, Stockholm, Swedish Institute for European Policy Issues.

Elgström O. (2007), Outsiders' Perceptions of the European Union in International Trade Negotiations. *Journal of Common Market Studies*, 45(4), 949-967.

Elgström O., Frennhoff Larsén M. (2010), Free to trade? Commission autonomy in the Economic Partnership Agreement negotiations. *Journal of European Public Policy*, 17(2), 205-223.

Elgström O., Jönsson C. (2000), Negotiation in the European Union: bargaining or problem-solving?. *Journal of European Public Policy*, 7(5), 684-704.

Elsig M. (2007a), The EU's Choice of Regulatory Venues for Trade Negotiations: A Tale of Agency Power?. *Journal of Common Market Studies*, 45(4), 927-948.

Elsig M. (2007b), *Delegation and Agency in EU Trade Policy Making: Bringing Brussels Back in*, NCCR working paper 2007/21, Geneva.

Emiliou N. (1996), The Allocation of Competence Between the EC and its Member States in the Sphere of External Relations, in Emiliou N., O'Keefe D. (eds.), *The European Union and World Trade Law*, Chichester, John Wiley & Sons, 31-45.

Epstein D., O'Halloran S. (1999), *Delegating Powers. A Transaction Cost Politics Approach to Policy Making Under Separate Powers*, New York, Cambridge University Press.

Falkner R. (2000), Regulating biotech trade: the Cartagena Protocol on Biosafety. *International Affairs*, 76(2), 299-313.

Falkner R. (2002), Negotiating the biosafety protocol: the international process, in Bail C., Falkner R., Marquard H. (eds.), *The Cartagena Protocol on Biosafety. Reconciling Trade in Biotechnology with Environment & Development?*, London, Royal Institute of International Affairs.

Falkner R. (2007), The political economy of 'normative power' Europe: EU environmental leadership in international biotechnology regulation. *Journal of European Public Policy*, 14(4), 507-526.

Feldmann L. (2004), Die strategische Umweltprüfung im Völkerrecht (SEA-Protokoll zur Espoo-Konvention), in Hendler R., Marburger P., Reinhart M., Schröder M. (eds.), *Die strategische Umweltprüfung (sog. Plan-UVP) als neues Instrument des Umweltsrecht*, Berlin, Erich Schmidt Verlag, 27-36.

Fernandez A. (2008), Change and Stability of the EU Institutional System: the Communitarization of the Council Presidency. *Journal of European Integration*, 30(5), 617-634.

Franchino F. (2007), *The Powers of the Union: Delegation in the EU*, New York, Cambridge University Press.

Frennhoff Larsén M. (2007), Trade Negotiations between the EU and South Africa: A Three-Level Game. *Journal of Common Market Studies*, 45(4), 857-881.

Frieden J. (2004), One Europe, One Vote?. *European Union Politics*, 5(2), 261-276.

Gilardi F. (2002), *Regulation through Independent Agencies in Western Europe: New Institutionalist Perspectives*, paper presented at the workshop 'Theories of Regulation', Oxford, Nuffield College.

Graff L. (2002), The precautionary principle, in Bail C., Falkner R., Marquard H. (eds.), *The Cartagena Protocol on Biosafety. Reconciling Trade in Biotechnology with Environment & Development?*, London, Royal Institute of International Affairs, 410-422.

Groenleer M., van Schaik L. (2007), United We Stand? The European Union's International Actorness in the Cases of the International Criminal Court and the Kyoto Protocol. *Journal of Common Market Studies*, 45(5), 969-998.

Gupta A. (2000), Governing Trade in Genetically Modified Organisms. The Cartagena Protocol on Biosafety. *Environment*, 42(4), 22-33.

Gupta J., Ringius L. (2001), The EU's Climate Leadership: Reconciling Ambition and Reality. *International Environmental Agreements: Politics, Law and Economics*, 1(2), 281-299.

Hall P., Taylor R. (1996), Political Science and the Three Institutionalisms. *Political Studies*, 44(4), 936-957.

Hawkins D., Jacoby W. (2006), How Agents Matter, in Hawkins D., Lake D., Nielson D., Tierney M. (eds.), *Delegation and Agency in International Organizations*, New York, Cambridge University Press, 199-228.

Hawkins D., Lake D., Nielson D., Tierney M. (2006), Delegation Under Anarchy: States, International Organizations, and Principal-Agent Theory, in Hawkins D., Lake D., Nielson D., Tierney M. (eds.), *Delegation and Agency in International Organizations*, New York, Cambridge University Press, 3-38.

Hayes-Renshaw F., Wallace H. (1997), *The Council of Ministers*, London, MacMillan Press.

Heliskoski J. (2000), Internal Struggle for International Presence: The Exercise of Voting Rights Within the FAO, in Dashwood A., Hillion C. (eds.), *The General Law of E.C. External Relations*, London, Sweet & Maxwell, 79-99.

Heliskoski J. (2001), *Mixed Agreements as a Technique for Organizing the International Relations of the European Community and its Member States*, Den Haag, Kluwer Law International.

Hodson D. (2009), Reforming EU economic governance: A view from (and on) the principal-agent approach. *Comparative European Politics*, 7(4), 455-475.

Hoffman A. (2002), A Conceptualization of Trust in International Relations. *European Journal of International Relations*, 8(3), 375-401.

Hooghe L. (2005), Several Roads Lead to International Norms, but Few Via International Socialization: A Case Study of the European Commission. *International Organization*, 59(4), 861-898.

Hug S., König T. (2002), In View of Ratification: Governmental Preferences and Domestic Constraints at the Amsterdam Intergovernmental Conference. *International Organization*, 56(2), 447-476.

Jäger J., O'Riordan T. (1996), The History of Climate Change Science and Politics, in O'Riordan T., Jäger J. (eds.), *Politics of Climate Change. A European Perspective*, London, Routledge, 1-31.

Jamal A. (1997), The United Nations Convention to Combat Desertification in those Countries Experiencing Serious Drought and/or Desertification, Particularly in Africa; Implementing Agenda 21. *Review of European Community and International Environmental Law*, 6(1), 1-6.

Joerges C., Neyer J. (1997), Transforming strategic interaction into deliberative problem-solving: European comitology in the foodstuffs sector. *Journal of European Public Policy*, 4(4), 609-625.

Johnson M. (1998), *European Community Trade Policy and the Article 133 Committee*, London, Royal Institute for International Affairs.

Jokela M. (2002), European Union as a Global Policy Actor: The Case of Desertification, in Biermann F., Brohm R., Dingwerth K. (eds.), *Proceedings of the 2001 Berlin Conference on the Human Dimensions of Global Environmental Change 'Global Environmental Change and the Nation State'*, Potsdam, Potsdam Institute for Climate Impact Research, 308-316.

Jupille J. (1999), The European Union and International Outcomes. *International Organization*, 53(2), 409-425.

Jupille J., Caporaso J. (1998), States, Agency, and Rules: The European Union in Global Environmental Politics, in Rhodes C. (ed.), *The European Union in the World Community*, London, Lynne Rienner Publishers, 213-229.

Kanie N. (2003), Leadership in Multilateral Negotiation and Domestic Policy: The Netherlands at the Kyoto Protocol Negotiation. *International Negotiation*, 8(2), 339-365.

Kassim H., Menon A. (2003), The principal-agent approach and the study of the European Union: promise unfulfilled?. *Journal of European Public Policy*, 10(1), 121-139.

Kay A. (2003), Path dependency and the CAP. *Journal of European Public Policy*, 10(3), 405-420.

Kelemen D. (2010), Globalizing European Union environmental policy. *Journal of European Public Policy*, 17(3): 335-349.

Keohane R. (1986), Reciprocity in international relations. *International Organization*, 40(1), 1-27.

Kerremans B. (1996a), Do Institutions Make a Difference? Non-Institutionalism, Neo-Institutionalism, and the Logic of Common Decision-Making in the European Union. *Governance*, 9(2), 217-240.

Kerremans B. (1996b), *Besluitvorming en integratie in de externe economische betrekkingen van de Europese Unie*, Brussels, Koninklijke Academie voor Wetenschappen, Letteren en Schone Kunsten van België.

Kerremans B. (2003), *Coordination on External Trade Policies*, Katholieke Universiteit Leuven, Leuven.

Kerremans B. (2004a), What Went Wrong in Cancun? A Principal-Agent View on the EU's Rationale Towards the Doha Development Round. *European Foreign Affairs Review*, 9(3), 363-393.

Kerremans B. (2004b), The European Commission and the EU Member States as Actors in the WTO Negotiating Process: Decision-Making between Scylla and Charibdis?, in Reinalda B., Verbeek B. (eds.), *Decision Making Within International Organizations*, London, Routledge, 45-58.

Kerremans B. (2006), Pro-Active Policy Entrepreneur or Risk Minimizer? A Principal-Agent Interpretation of the EU's Role in the WTO, in Elgström O., Smith M. (eds.), *The European Union's Roles in International Politics*, Oxford, Routledge, 172-188.

Kiewiet D., McCubbins M. (1991), *The Logic of Delegation. Congressional Parties and the Appropriations Process,* Chicago, University of Chicago Press.

Kirsop B. (2002), The Cartagena (Biosafety) Protocol. *Journal of Commercial Biotechnology*, 8(3), 214-218.

Koenig-Archibugi M. (2004), Explaining Government Preferences for Institutional Change in EU Foreign and Security Policy. *International Organization*, 58(1), 137-174.

Kostadinova P. (2003), *Membership of the European Union: necessary and sufficient conditions (Boolean analysis based on the European Commission reports on Eastern Europe)*, paper prepared for the Annual Meeting of the American Political Science Association, Chicago.

Koutrakos P. (2006), *EU International Relations Law*, Portland, Hart Publishing.

Kummer K. (1999), Prior Informed Consent for Chemicals in International Trade: the 1998 Rotterdam Convention. *Review of European Community and International Environmental Law*, 8(3), 323-330.

Lacasta N., Dessai S., Powroslo E. (2002), *Consensus among many voices: articulating the European Union's position on Climate Change*, San Francisco, Golden Gate University, http://www.uea.ac.uk/ ~e120782/papers/consensus. html.

Lamy P. (2002), *Europe's Role in Global Governance: The Way Ahead*, speech at Humbolt University, Berlin.

Langlet D. (2003), Prior Informed Consent for Hazardous Chemicals Trade – Implementation in EC Law. *European Environmental Law Review*, 12(11), 292-308.

Leal-Arcas R. (2001), The European Community and Mixed Agreements. *European Foreign Affairs Review*, 6(4), 483-513.

Leal-Arcas R. (2003), United we Stand, Divided we Fall. The European Community and its Member States in the WTO Forum: towards greater Cooperation on Issues of Shared Competence?. *European Political Economy Review*, 1(1), 65-79.

Leal-Arcas R. (2004), *The EC in the WTO: The three-level game of decision-making. What multilateralism can learn from regionalism*, s.l., European Integration online Papers, http://eiop.or.at/eiop/texte/2004-014a.htm.

Leefmans P. (1998), *Externe milieubevoegdheden. Communautairrechtelijke grenzen aan externe milieubevoegdheden van de EG-lidstaten*, Deventer, Kluwer.

Lenaerts K., Van Nuffel P., Bray R. (2005), *Constitutional Law of the European Union*, London, Sweet & Maxwell.

Lewis J. (1998), Is the 'Hard Bargaining' Image of the Council Misleading? The Committee of Permanent Representatives and the Local Elections Directive. *Journal of Common Market Studies*, 36(4), 479-504.

Lewis J. (2000), The methods of community in EU decision-making and administrative rivalry in the Council's infrastructure. *Journal of European Public Policy*, 7(2), 261-289.

Lewis J. (2003a), Informal integration and the supranational construction of the Council. *Journal of European Public Policy*, 10(6), 996-1019.

Lewis J. (2003b), Institutional Environments and Everyday EU Decision Making. Rationalist or Constructivist? in *Comparative Political Studies*, 36(1/2), 97-124.

Lewis J. (2005), The Janus Face of Brussels: Socialization and Everyday Decision Making in the European Union. *International Organization*, 59(4), 937-971.

Liberatore A. (1997), The European Union: bridging domestic and international environmental policy-making, in Schreurs M., Economy E. (eds.), *The*

Internationalization of Environmental Protection, New York, Cambridge University Press, 188-212.

Lijphart A. (1975), The Comparable-Cases strategy in Comparative Research. *Comparative Political Studies*, 8(2), 158-177.

Lutz K., Boye P., Haupt H. (2000), Zur Entstehungsgeschichte des AEWA. *Schriftenr. Landschaftspflege Naturschutz*, 60(–), 7-12.

Lyne M., Nielson D., Tierney M. (2003), *A Problem of Principals: Common Agency and Social Lending at the Multilateral Development Banks*, meeting on Delegation to International Organizations, Del Mar.

Macleod I., Hendry I., Hyett S. (1998), *The External Relations of the European Communities. A Manual of Law and Practice*, Oxford, Clarendon Press.

Macrory R., Hession M. (1996), The European Community and Climate Change. The role of law and legal competence, in O'Riordan T., Jäger J. (eds.), *Politics of Climate Change. A European Perspective*, London, Routledge, 106-154.

Maher I., Billiet S., Hodson D. (2009), The principal-agent approach to EU studies: Apply liberally but handle with care. *Comparative European Politics*, 7(4), 409-413.

Martin L., Simmons B. (1998), Theories and Empirical Studies of International Institutions. *International Organization*, 52(4), 729-757.

Maskin E., Tirole J. (1990), The Principal-Agent Relationship with an Informed Principal: The Case of Private Values. *Econometrica*, 58(2), 379-409.

Maurer A., Kietz D., Völkel C. (2005), Interinstitutional Agreements in the CFSP: Parliamentarization through the Back Door?. *European Foreign Affairs Review*, 10(4), 175-195.

McGoldrick D. (1997), *International Relations Law of the European Union*, London, Longman.

McNamara K. (2002), Rational Fictions: Central Bank Independence and the Social Logic of Delegation. *West European Politics*, 25(1), 47-76.

Menon A. (2003), Member States and International Institutions: Institutionalizing Intergovernmentalism in the European Union. *Comparative European Politics*, 1(2), 171-201.

Metcalfe D. (1998), Leadership in European Union Negotiations: The Presidency of the Council. *International Negotiation*, 3(3), 413-434.

Meunier S. (1998), Divided but United: European Trade Policy Integration and EC-U.S. Agricultural Negotiations in the Uruguay Round, in Rhodes C. (ed.), *The European Union in the World Community*, London, Lynne Rienner Publishers, 193-211.

Meunier S. (2000), What Single Voice? European Institutions and EU-U.S. Trade Negotiations. *International Organization*, 54(1), 103-135.

Meunier S. (2005), *Trading Voices. The European Union in International Commercial Negotiations*, Princeton, Princeton University Press.

Meunier S. (2007), Managing Globalization? The EU in International Trade Negotiations. *Journal of Common Market Studies*, 45(4), 905-926.

Meunier S., Nicolaïdis C. (2006), The European Union as a conflicted trade power. *Journal of European Public Policy*, 13(6), 906-925.

Meunier S., Nicolaïdis K. (2000), EU Trade Policy: The Exclusive versus Shared Competence Debate, in Cowles M., Smith M. (eds.), *The State of the European Union. Risks, Reform, Resistance, and Revival*, New York, Oxford University Press, 325-346.

Milner H. (1997), *Interests, institutions and information. Domestic politics and international relations*, Princeton New Jersey, Princeton University Press.

Mo J. (1995), Domestic Institutions and International Bargaining: The Role of Agent Veto in Two-Level Games. *American Political Science Review*, 89(4), 914-924.

Moravcsik A. (1993), Introduction. Integrating international and domestic theories of international bargaining. Evans P., Jacobson H., Putnam R. (eds.), *International bargaining and domestic politics. Double-edged diplomacy*, Los Angeles, University of California Press, 3-42.

Moser P., Schneider G. (2000), Rational Choice and the Governance of the European Union: An Introduction, in Moser P., Schneider G., Kirchgässner G. (eds.), *Decision Rules in the European Union. A Rational Choice Perspective*, London, MacMillan Press, 1-15.

Murphy A. (2000), In the maelstrom of change. The Article 133 Committee in the governance of external economic policy. Christiansen T., Kirchner E. (eds.), *Committee Governance in the European Union*, Manchester, Manchester University Press, 98-114.

Najam A. (2004), Dynamics of the Southern Collective: Developing Countries in Desertification Negotiations. *Global Environmental Politics*, 4(3), 128-154.

Newell P. (1998), Who 'CoPed' out in Kyoto? An Assessment of the Third Conference of the Parties to the Framework Convention on Climate Change. *Environmental Politics*, 7(2), 153-159.

Nicolaïdis K. (1999), Minimizing Agency Costs in Two-Level Games. Lessons From the Trade Authority Controversies in the United States and the European Union, in Mnookin R., Susskind L. (eds.), *Negotiating on behalf of others*, London, Sage Publications, 87-126.

Nielson D., Tierney M. (2003), Delegation to International Organisations: Agency Theory and World Bank Environmental Reform. *International Organization*, 57(2), 241-276.

Niemann A. (2004), Between communicative action and strategic action: the Article 133 Committee and the negotiations on the WTO Basic Telecommunications Services Agreement. *Journal of European Public Policy*, 11(3), 379-407.

Oberthür S., Ott H. (1999), *The Kyoto Protocol. International Climate Change Policy for the 21st Century*, Berlin, Springer-Verlag.

Oberthür S., Roche Kelly C. (2008), EU Leadership in International Climate Policy: Achievements and Challenges. *The International Spectator*, 45(3), 35-50.

Ott H. (2001), Climate change: an important foreign policy issue. *International Affairs*, 77(2), 277-296.

Pajala A., Widgrén M. (2004), A Priori versus Empirical Voting Power in the EU Council of Ministers. *European Union Politics*, 5(1), 73-97.

Palerm J. (1999), Public Participation in Environmental Decision Making: Examining the Aarhus Convention. *Journal of Environmental Assessment Policy and Management*, 1(2), 229-244.

Pallemaerts M. (2004), De Europese Gemeenschap als Verdragsluitende Partij bij het Protocol van Kyoto, in Maes F. (ed.), *Verhandelbare emissierechten als klimaatbeleidsinstrument. L'échange des droits de pollution comme instrument de gestion du climat*, Bruges, Die Keure, 6-27.

Pattersson L. (1997), Agricultural policy reform in the European Community: a three-level-game analysis. *International Organization*, 51(1), 135-165.

Peterson J., Bomberg E. (1999), *Decision-making in the European Union*, London, MacMillan Press.

Pierson P. (2000), Increasing Returns, Path Dependence, and the Study of Politics. *American Political Science Review*, 94(2), 251-267.

Pinholt K. (2004), *Influence through arguments? A study of the Commission's influence on the climate change negotiations*, Oslo, Arena Report 03/2004.

Pocar F. (2002), The Decision-Making Process of the European Community in External Relations, in Cannizzaro E. (ed.), *The European Union as an Actor in International Relations*, Den Haag, Kluwer Law International, 3-16.

Pollack M. (2002), Learning from the Americanists (Again), Theory and Method in the Study of Delegation. *West European Politics*, 25(1), 200-219.

Pollack M. (2003a), *The Engines of European Integration. Delegation, Agency, and Agenda Setting in the EU*, New York, Oxford University Press.

Pollack M. (2003b), Control Mechanism or Deliberative Democracy? Two Images of Comitology. *Comparative Political Studies*, 36(1/2), 125-155.

Pollack M. (2004), The New Institutionalisms and European Integration, in Wiener A., Diez T. (eds.), *European Integration Theory*, New York, Oxford University Press, 137-156.

Putnam R. (1988), Diplomacy and domestic policies: the logic of two-level games. *International Organization*, 42(3), 427-460.

Raffensperger C., Tickner J. (1999), *Protecting Public Health and the Environment*, Washington DC, Island Press.

Ragin C. (1987), *The Comparative Method. Moving Beyond Qualitative and Quantitative Strategies*, Los Angeles, University of California Press.

Ragin C., Berg-Schlosser D., De Meur G. (1996), Political Methodology: Qualitative Methods, in Goodin R., Klingemann H. (eds.), *A New Handbook of Political Science*, New York, Oxford University Press, 749-768.

Rasmussen A. (2000), *Institutional Games Rational Actors Play. The empowering of the European Parliament*, s.l., European Integration online Papers, http://eiop.or.at/eiop/texte/2000-001a.htm.

Rasmussen A. (2005), EU Conciliation Delegates: Responsible or Runaway Agents?. *West European Politics*, 28(5), 1015-1034.

Rhinard M., Kaeding M. (2006), The International Bargaining Power of the European Union in 'Mixed' Competence Negotiations: The Case of the 2000 Cartagena Protocol on Biosafety. *Journal of Common Market Studies*, 44(5), 1023-1050.

Rihoux B. (2003), Bridging the Gap between the Qualitative and Quantitative Worlds? A Retrospective and Prospective View on Qualitative Comparative Analysis. *Field Methods*, 15(4), 351-365.

Rihoux B., Ragin C. (2004), *Qualitative Comparative Analysis (QCA), State of the Art and Prospects*, Chicago, paper presented at annual meeting of the American Political Science Association, http://www.allacademic.com/meta/p61198_index.html.

Rittberger B., Schimmelfennig F. (2005), *The Constitutionalization of the European Union: A Qualitative Comparative Analysis*, Oslo, paper presented at Arena, http://www.arena.uio.no/events/ seminarpapers/Rittberger_Nov05.pdf.

Rodenhoff V. (2002), The Aarhus Convention and its Implications for the "Institutions" of the European Community. *Review of European Community and International Environmental Law*, 11(3), 343-357.

Salacuse J. (1999), Law and Power in Agency Relationships, in Mnookin R., Susskind L. (eds.), *Negotiating on behalf of others*, London, Sage Publications, 157-175.

Sandholtz W. (1996), Membership Matters: Limits of the Functional Approach to European Institutions. *Journal of Common Market Studies*, 34(3), 403-429.

Sartori G. (1994), Bien comparer, mal comparer. *Revue Internationale de Politique Comparée*, 1(1), 19-36.

Sbragia A. (1998), Institution-Building from below and above: The European Community in Global Environmental Politics, in Sandholz W., Stone Sweet A. (eds.), *European Integration and Supranational Governance*, New York, Oxford University Press, 283-303.

Schalk J., Torenvlied R., Weesie J., Stokman F. (2007), The Power of the Presidency in EU Council Decision-making. *European Union Politics*, 8(2), 229-250.

Schmidt J. (2008), Why Europe Leads on Climate Change. *Survival*, 50(4), 83-96.

Schmidt S. (2000), Only an Agenda Setter? The European Commission's Power over the Council of Ministers. *European Union Politics*, 1(1), 37-61.

Schout A., Vanhoonacker S. (2006), Evaluating Presidencies of the Council of the EU: Revisiting Nice. *Journal of Common Market Studies*, 44(5), 1051-1077.

Schreurs M., Tiberghien Y. (2007), Multi-Level Reinforcement: Explaining European Union Leadership in Climate Change Mitigation. *Global Environmental Politics*, 7(4), 19-46.

Shapiro S. (2005), Agency Theory. *Annual Review of Sociology*, 31(–), 263-284.

Sjöstedt G. (1998), The EU Negotiates Climate Change. External Performance and Internal Structural Change. *Cooperation and conflict*, 33(3), 227-256.

Smith K. (2006), The European Union, Human Rights and the United Nations, in Smith K., Laatikainen K. (eds.), *Intersecting Multilateralisms. The European Union and the United Nations*, New York, Palgrave Macmillan, 154-174.

Smyrl M. (1998), When (and How) Do the Commission's Preferences Matter?. *Journal of Common Market Studies*, 36(1), 79-99.

Svensson A. (2000), *In the Service of the European Union. The Role of the Presidency in Negotiating the Amsterdam Treaty 1995-97*, Uppsala, Uppsala University.

Tallberg J. (2002), Delegation to Supranational Institutions: Why, How, and with What Consequences?. *West European Politics*, 25(1), 23-46.

Tallberg J. (2003), The agenda-shaping powers of the EU Council Presidency. *Journal of European Public Policy*, 10(1), 1-19.

Tallberg J. (2006), *Leadership and Negotiation in the European Union*, New York, Cambridge University Press.

Thieme D. (2001), European Community External Relations in the Field of the Environment. *European Environmental Law Review*, 11(8-9), 252-264.

Tiberghien Y., Starrs S. (2004), *The EU as Global Trouble-Maker in Chief: A Political Analysis of EU Regulations and EU Global Leadership in the Field of Genetically Modified Organisms*, paper presented at the Conference for Europeanists, Chicago.

Toulmin C. (1995), Combating desertification by conventional means. *Global Environmental Change*, 5(5), 455-457.

Tsebelis G. (2002), *Veto Players. How Political Institutions Work*, Princeton, Princeton University Press.

van Calster G., Lee M. (2002), European Case Law Report September 2001 – March 2002. *Review of European Community and International Environmental Law*, 11(2), 235-245.

van den Hoven A. (2004), Assuming Leadership in Multilateral Economic Institutions: The EU's "Development Round" Discourse and Strategy. *West European Politics*, 27(2), 256-283.

Van Schendelen M. (1996), 'The Council Decides': Does the Council Decide?. *Journal of Common Market Studies*, 34(4), 531-548.

Vanden Bilcke C. (2002), The Stockholm Convention on Persistent Organic Pollutants. *Review of European Community & International Environmental Law*, 11(3), 328-342.

Veinla H., Relve K. (2005), Influence of the Aarhus Convention on Access to Justice in Environmental Matters in Estonia. *European Environmental Law Review*, 14(12), 326-331.

Verwey D. (2004), *The European Community, the European Union and the International Law of Treaties*, Den Haag, TMC Asser Press.

Vogler J. (1999), The European Union as an Actor in International Environmental Politics. *Environmental Politics*, 8(3), 24-48.

Vogler J. (2005), The European contribution to global environmental governance. *International Affairs*, 81(4), 835-850.

Vogler J. (2009), Climate change and EU foreign policy: The negotiation of burden sharing. *International Politics*, 46(4), 469-490.

Wallström M. (2002), European Commission, in Bail C., Falkner R., Marquard H. (eds.), *The Cartagena Protocol on Biosafety. Reconciling Trade in Biotechnology with Environment & Development?*, London, Royal Institute of International Affairs, 244-250.

Waterman R., Meier K. (1998), Principal-Agent Models: An Expansion?. *Journal of Public Administration Research and Theory*, 8(2), 173-202.

Winfield M. (2000), *Reflections on the Biosafety Protocol Negotiations in Montreal*, s.l., Bio-Tech Info, http://www.biotech-info.net/BSP_reflections.html.

Woll C. (2006), The road to external representation: the European Commission's activism in international air transport. *Journal of European Public Policy*, 13(1), 52-69.

Worsham J., Eisner M., Ringquist E. (1997), Assessing the Assumptions. A Critical Analysis of Agency Theory. *Administration & Society*, 28a(4), 419-440.

Yamin F. (1998), The Kyoto Protocol: Origins, Assessment and Future Challenges. *Review of European Community and International Environmental Law*, 7(2), 113-127.

Young A. (2003), What game? By which rules? Adapting and flexibility in the EC's foreign economic policy, in Knodt M., Princen S. (eds.), *Understanding the European Union's External Relations*, London, Routledge, 54-71.

Zito A. (2005), The European Union as an Environmental Leader in a Global Environment. *Globalizations*, 2(3), 363-375.

Index

GLOBAL ENVIRONMENTAL GOVERNANCE SERIES

Full series list

Agricultural Policy Reform
Politics and Process in the EU and US
in the 1990s
Wayne Moyer and Tim Josling

Linking Trade, Environment, and
Social Cohesion
NAFTA Experiences, Global Challenges
*Edited by John J. Kirton and
Virginia W. Maclaren*

International Equity and Global
Environmental Politics
Power and Principles in US Foreign Policy
Paul G. Harris